COMMENTS FROM THE FIELD ABOUT
WE REASON & WE PROVE FOR ALL MATHEMATICS

"Simply stated, this book is a *must-have* for preservice and inservice mathematics teachers and teacher leaders who are looking to enhance their understanding of how reasoning-and-proving are critical processes for increasing proficiency across *all* mathematics content domains. This expert author team has illustrated a clear vision and plan, supported by key strategies and exceptional tools, for guiding teacher teams as they help their students learn how to make conjectures and develop and judge the effectiveness of their arguments and proofs. This book is an exceptionally useful and timely resource for schools and districts that are looking to connect and deepen their professional focus with the Effective Mathematics Teaching Practices (NCTM, 2014) and other evidence-based practices."

Jonathan (Jon) Wray
Coordinator of Secondary Mathematics, Howard County Public Schools (MD)
Board of Directors, National Council of Teachers of Mathematics

"Reasoning-and-proving are central to investigating ideas, solving problems, and establishing mathematics knowledge at all levels. Built around rich classroom cases, this book provides research-supported frameworks and practical resources for teachers to deepen their understanding and develop practices to aid students in reasoning-and-proving as powerful mathematical thinkers."

Daniel Heck
Vice President of Horizon Research, Inc.

"*We Reason & We Prove for ALL Mathematics* provides an enlightening and engaging examination of reasoning-and-proving in secondary mathematics classrooms. Filled with carefully designed tasks and task sequences, along with illustrative classroom cases, it clearly articulates the nature of reasoning-and-proving, what students need to know and understand about it, and how teachers can support this learning. The thought-provoking discussion questions and recommended classroom activities support readers' implementation of reasoning-and-proving activities into their own classrooms. *We Reason & We Prove for ALL Mathematics* is an outstanding resource for practice-based learning on this essential component of mathematics learning. I recommend it most highly."

Diane J. Briars, PhD
Mathematics Education Consultant
Past President, National Council of Teachers of Mathematics (NCTM)

We Reason & We Prove for ALL Mathematics

We Reason & We Prove for ALL Mathematics

Building Students' Critical Thinking, Grades 6–12

Fran Arbaugh
Margaret (Peg) Smith
Justin Boyle
Gabriel J. Stylianides
Michael Steele

CORWIN Mathematics

For information:

Corwin
A SAGE Company
2455 Teller Road
Thousand Oaks, California 91320
(800) 233-9936
www.corwin.com

SAGE Publications Ltd.
1 Oliver's Yard
55 City Road
London, EC1Y 1SP
United Kingdom

SAGE Publications India Pvt. Ltd.
B 1/I 1 Mohan Cooperative Industrial Area
Mathura Road, New Delhi 110 044
India

SAGE Publications Asia-Pacific Pte. Ltd.
3 Church Street
#10–04 Samsung Hub
Singapore 049483

Program Manager, Mathematics: Erin Null
Associate Editor: Julie Nemer
Editorial Assistant: Jessica Vidal
Production Editor: Tori Mirsadjadi
Copy Editor: Liann Lech
Typesetter: C&M Digitals (P) Ltd.
Proofreader: Jen Grubba
Indexer: William Ragsdale
Cover and Interior Designer: Scott Van Atta
Marketing Manager: Margaret O'Connor

Library of Congress Cataloging-in-Publication Data

Names: Arbaugh, Fran, author.

Title: We reason & we prove for all mathematics : building students' critical thinking, grades 6–12 / Fran Arbaugh [and four others].

Other titles: We reason and we prove for all mathematics
Description: Thousand Oaks, California : Corwin, a Sage Company, [2018] | Includes bibliographical references and index.

Identifiers: LCCN 2018017120 | ISBN 9781506378190 (pbk. : alk. paper)

Subjects: LCSH: Mathematics–Study and teaching (Elementary) | Mathematics–Study and teaching (Middle school) | Mathematics–Study and teaching (Secondary) | Curriculum planning.

Classification: LCC QA11.2 .W4225 2018 | DDC 510.71/2–dc23 LC record available at https://lccn.loc.gov/2018017120

Contents

 For bonus materials and resources, visit the companion website at **resources.corwin.com/reasonandprove**

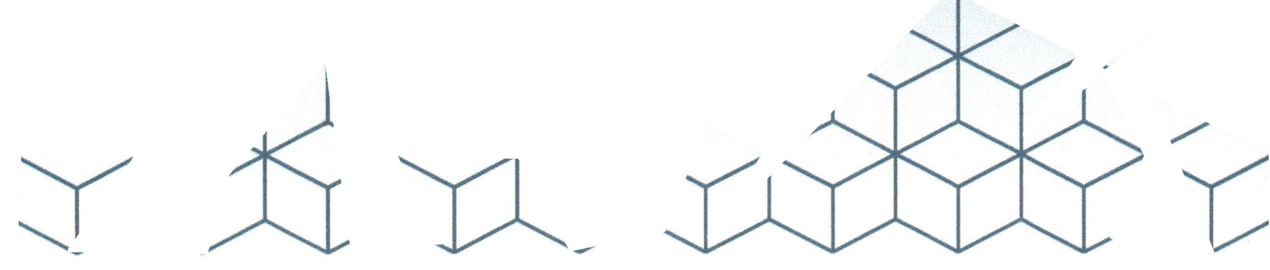

Preface

What do you think of when you hear the word *proof*? It may bring to mind a set of statements and reasons in two columns that you encountered in high school geometry (e.g., see Figure P.1) or perhaps a number theory proof by mathematical induction from a college course (e.g., see Figure P.2). Perhaps when you hear the word *proof* you think of something that is hard to do, or a topic not appropriate for most secondary students. If you have any of these thoughts about proof, you are not alone. For example, in a group of 192 pre-

FIGURE P.1 Geometric proof in two columns.

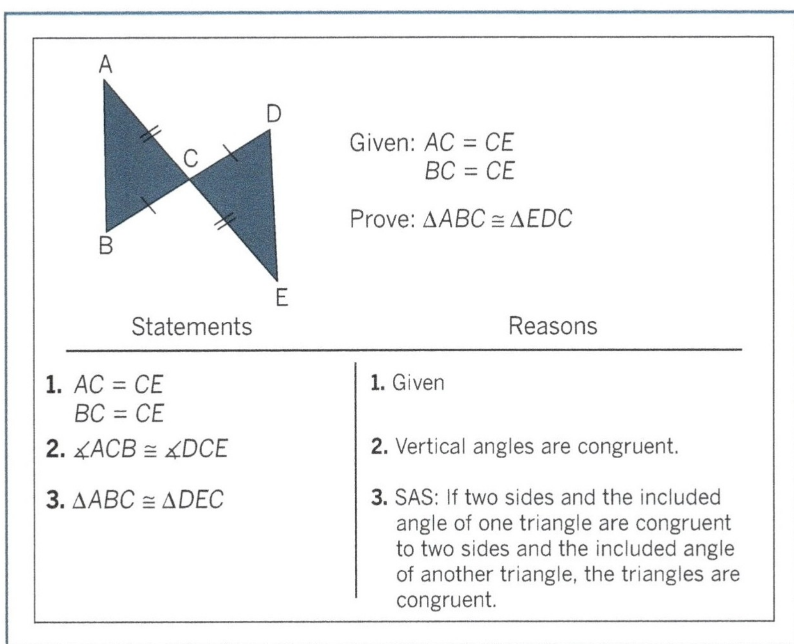

Problem: For any natural number *n*, *n³* + *2n* is divisible by *3*.

Proof:
Basis Step: If *n = 0*, then *n³* + *2n = 0³* + *2*0 = 0*. So it is divisible by *3*.
Induction: Assume that for an arbitrary natural number *n*,
n³ + *2n* is divisible by *3*. -------- *Induction Hypothesis*
To prove this for *n* + *1*, first try to express *(n* + *1)³* + *2(n* + *1)* in terms of
n³ + *2n* and use the induction hypothesis.

$$(n + 1)^3 + 2(n + 1) = (n^3 + 3n^2 + 3n + 1) + (2n + 2)$$
$$= (n^3 + 2n) + (3n^2 + 3n + 3)$$
$$= (n^3 + 2n) + 3(n^2 + n + 1)$$

which is divisible by *3*, because *(n³* + *2n)* is divisible by *3* by the induction
hypothesis.

End of Proof.

Source: http://www.cs.odu.edu/~cs381/cs381content/induction/example5/example5.html

service mathematics teachers, 18% had negative opinions toward mathematical proof and provided responses that were classified as "proof is hard to understand," "proof is unnecessary," and "proof is the nightmare of the students" (Ersen, 2016). Knuth (2002b) found that the secondary teachers he interviewed saw proof as "an appropriate idea only for those students enrolled in advanced mathematics classes and for those students who will most likely be pursuing mathematics-related majors in college" (pp. 73–74).

The purpose of this book is to dispel some of the common conceptions regarding what proof is, how it can be written, and when and for whom it is appropriate. Alan Schoenfeld, a prominent mathematics educator, stated,

> Proof is not a thing separable from mathematics, as it appears to be in our curricula; it is an essential component of doing, communicating, and recording mathematics. And I believe it can be embedded in our curricula, at all levels. (1994, p. 76)

We agree with Schoenfeld's claim, and in the chapters that follow, we will provide guidance about how to make reasoning-and-proving a reality in your classroom.

WHAT THIS BOOK IS ABOUT

The content of this book is based on the importance of mathematical reasoning-and-proving in middle and high school mathematics—processes that can and do enhance students' understandings of the math that they are learning and help them to retain those understandings over time. Much of what we know from over 40 years of careful and rigorous research about learning and understanding mathematics is that kids of all ages benefit in many ways when they are required to make sense of the mathematics they are learning—when they engage in sense-making as a regular activity in math class. We believe that a powerful way to engage students in sense-making is to engage them in reasoning-and-proving activities.

What does that kind of instruction look like? How do you teach in a way that supports reasoning-and-proving in your classroom? This book will help to enhance your own understanding of reasoning-and-proving, and provide practical teaching practices to help your students reason-and-prove in your middle and high school mathematics classrooms.

Why learn more about how to incorporate reasoning-and-proving into your mathematics instruction? One reason is that the field of mathematics education has long emphasized the importance of reasoning and sense-making in K–12 mathematics for the learning of mathematics with understanding. A common thread that has run through standards and policy documents (e.g., NCTM, 1989, 1991, 1995, 2000) over the past three decades, as well as in professional publications about the teaching of mathematics (e.g., NCTM, 2006, 2009, 2014), is that they advocate for the necessity of learning environments that encourage and support mathematical reasoning-and-proving. This thread continued with the publication of the *Common Core State Standards–Mathematics* (CCSSM) (National Governors Association & Council of Chief State School Officers, 2010), which also advocates for these types of learning environments through articulating the eight Standards for Mathematical Practice (more on those in Chapter 1).

But even with repeated calls for incorporating reasoning and sense-making into K–12 mathematics learning environments, research has documented that students' opportunities to engage in environments that encourage reasoning, sense-making, and proving have been quite limited (Horizon Research, Inc., 2013; Weiss, Pasley, Smith, Banilower, & Heck, 2003). If we want our students to be able to make sense of the mathematics they are learning, then we believe that incorporating reasoning-and-proving activities into middle and high school mathematics has the potential for strongly supporting students' understanding of important mathematics.

One of the challenges of realizing the vision of reasoning-and-proving set forth in the documents referenced above is that many mathematics teachers have not had the opportunity to engage in reasoning-and-proving activities as mathematics learners or to consider how to incorporate reasoning-and-proving into their mathematics teaching. The vast majority of us never learned mathematics in this way, making it difficult to envision the kinds of mathematics learning environments that support our students to be mathematical thinkers.

If you are one of those teachers, this book is for you.

THE PROJECT BEHIND THIS BOOK

As a group of former mathematics teachers and current university-level mathematics teacher educators, we were concerned about the state of reasoning-and-proving that had been documented by research conducted in U.S. mathematics classrooms. As a result of this concern, we worked together to develop and propose a research and materials development project that focused on helping mathematics teachers learn more about reasoning-and-proving and how these processes play out in classrooms. As a result of this proposal, the National Science Foundation (NSF) awarded us a grant that supported the development of the contents of this book under the auspices of a project titled *Cases of Reasoning-and-Proving (CORP) in Secondary Mathematics* (NSF #DRL 0732798). Fran Arbaugh and Margaret (Peg) Smith, the first two authors of this book, served as co-directors of the project with the remaining co-authors, Justin Boyle, Gabriel J. Stylianides, and Michael Steele, helping to develop the materials. Other mathematics educators contributed to the development of the materials in different ways, and we express appreciation for their contributions in the book's acknowledgements.

During development, we piloted the CORP materials with both preservice and practicing middle and high school mathematics teachers and revised them based on what we learned from those pilots. So, the learning activities that form the core of this book are tried and tested with hundreds of teachers. More formal evaluation of the materials (e.g., Boyle, 2012; Karunakaran, Freeburn, Konuk, & Arbaugh, 2014) has confirmed that teachers learned a great deal about reasoning-and-proving from interacting with the activities. In short, we have found that our approach for learning about reasoning-and-proving, as well as how to implement reasoning-and-proving activities in mathematics classrooms, is effective. We now want to give more teachers access to the opportunity to learn about reasoning-and-proving through the publication of this book.

HOW TO USE THIS BOOK

The primary audience of this book is practicing middle and high school mathematics teachers and those who are studying to be middle and high school mathematics teachers. Teachers can learn from the book by engaging with it in a number of different ways:

- **Individuals** can use it for independent study, working through the book on their own.
- **Small groups of teachers** can use this book as the basis for a book study in a professional learning community, working through the book together and engaging in discussions about the content.
- **Professional developers** can use this book to plan and implement professional development focused on enhancing middle and high school teachers' knowledge and teaching practices focused on reasoning-and-proving.
- **Mathematics teacher educators** can use this book in planning and implementing university-level coursework in mathematics education.

We wrote this book in a way that encourages readers to engage actively in learning activities about reasoning-and-proving because we believe that the best learning occurs through doing and reflecting. Smith (2001) described engagement in such activities as "practice-based professional development." We believe that simply reading through the book without doing the activities will limit what you learn. In other words, you will get the most out of the book if you **stop and engage** as you work through the book.

Across the chapters, there are mathematical tasks for you to do, sets of student work for you to analyze, narrative cases of mathematics classrooms for you to examine and reflect upon, and "Pause and Consider" prompts to support you in reflecting on what you are learning. We based the narrative cases on events that occurred in real mathematics classrooms, and while not written verbatim from classroom audio- and video-recordings, we have represented with fidelity the heart of what happened in the classrooms. The examples of student work are from real students, many of which have been rewritten for the purpose of making them more readable.

We have clearly indicated those learning activities and reflective prompts within the chapters and have included some blank pages in the back of the book for you to use for recording your reflections or making teaching notes along the way. It is likely, though, that you will need a separate notebook where you can complete the mathematical tasks that are part of many activities. The narrative cases are included in the book as appendixes, and there are several documents on the companion website that you can download and use along the way—look for the website indicator in the margin to help you know what is available on the website.

The book is organized into seven chapters, and the content builds from chapter to chapter—the activities in later chapters depend on you having worked through activities in earlier chapters. So, it is important that you work through the chapters sequentially. Here is a brief preview of each chapter:

- In **Chapter 1: Setting the Stage**, you will reflect on your current conceptions of proof and the role of proof in middle and high school mathematics and read more about the background for the book. We also provide information about our conceptions of reasoning-and-proving that undergird the contents of the book, including why we use a hyphenated version of the phrase.
- In **Chapter 2: Convincing Students That Proof Matters**, you will engage in a series of mathematical tasks that you can use with your students to convince them that the use of numerous examples is not enough to prove a mathematical conjecture. Then you will consider the implementation of the series of tasks in two different mathematics classrooms through reading and analyzing narrative cases.
- In **Chapter 3: Exploring the Nature of Reasoning-and-Proving**, you will develop a set of criteria that can be used to judge when a mathematical argument counts as proof. Then

you will learn about a framework for reasoning-and-proving that Gabriel J. Stylianides (one of the authors of this book) developed and consider how that framework can be used to organize reasoning-and-proving activities in your classroom.

- In **Chapter 4: Helping Students Develop the Capacity to Reason-and-Prove,** you will consider four effective teaching practices that you can use in your classroom to support your students' engagement in reasoning-and-proving activities. Then you will visit two mathematics classrooms, through narrative cases, and analyze how those teaching practices are implemented with students.

- In **Chapter 5: Modifying Tasks to Increase Reasoning-and-Proving Potential,** you will learn about strategies for modifying tasks from curriculum materials so that they better support students to engage in reasoning-and-proving. In addition, you will consider how establishing clear learning goals supports implementation of reasoning-and-proving activities with your students.

- In **Chapter 6: Using Context to Engage in Reasoning-and-Proving,** you will focus on how context supports students to reason-and-prove as you solve a reasoning-and-proving task, analyze a set of student responses to that task, and then analyze two narrative cases captured in mathematics classrooms where the task was implemented with high school students.

- In **Chapter 7: Pulling It All Together,** you will reflect back on the content of the book, considering key ideas presented in Chapters 1–6 and important tools that support engaging in reasoning-and-proving as well as the teaching of reasoning-and-proving.

We hope that this book provides a valuable learning opportunity for you. You will certainly broaden and deepen your current understandings of reasoning-and-proving *and* how to create learning environments that help students to enhance their capacities to reason-and-prove.

Fran Arbaugh
Margaret (Peg) Smith
Justin Boyle
Gabriel J. Stylianides
Michael Steele

Acknowledgments

The activities that appear in this book were developed under the auspices of the *Cases of Reasoning-and-Proving (CORP) in Secondary Mathematics* project that was funded by the National Science Foundation under award no. DRL 0732798. The views expressed herein are those of the authors and do not necessarily represent the views of the supporting agency.

Over the course of the project, many colleagues—beyond the authors of this book—contributed to the CORP project in important ways. The expertise, insights, commitment, and support we received from our collaborators were invaluable. Specifically, we acknowledge the thoughtful contributions of the following individuals:

- James Greeno and Gaea Leinhardt for their input on the conceptualization of the project and input throughout the development process;
- Amy Hillen for assistance in designing and creating early versions of materials;
- Doctoral students Cynthia Taylor, Nursen Konuk, Ben Freeburn, Shiv Karunakaran, Michelle Switala, and Adam Vrabel for assistance in designing materials and/or designing, teaching, and studying courses in which the materials were used;
- A panel of expert advisors who provided valuable feedback at key points in the materials development process—Hyman Bass (University of Michigan); Guershon Harel (University of California, San Diego); Eric Knuth (University of Texas, Austin); William McCallum (University of Arizona); Sharon Senk (Michigan State University); and Edward Silver (University of Michigan);

- Evelyn Gordon and Daniel Heck from Horizon Research, who provided yearly evaluation reports that helped us in making midcourse adjustments in the materials; and
- The teachers who opened their classrooms to us so that we could all learn from their efforts to increase reasoning-and-proving opportunities for their students—Gina Burrows, Charlie Sanders, Nancy Edwards, Vicky Mansfield, Natalie Boyer, and Calvin Jensen.

PUBLISHER'S ACKNOWLEDGMENTS

Julie Conrad
Mathematics Coordinator and Educational Consultant
Vermont

Kenneth Davis
Mathematics Coordinator
Madison Metropolitan School District
Madison, Wisconsin

Fred Dillon
Mathematics Coach
Ideastream
Cleveland, Ohio

Diana L. Kasbaum
ASSM Presidential Awards Coordinator and Independent Consultant
2001–2015 Mathematics Education Consultant for WI Department
of Instruction
Wisconsin

Jenny Salls
Mathematics Teacher, Secondary
Clayton Pre-AP Academy
Reno, Nevada

John W. Staley
Past President, NCSM (National Council of Supervisors
of Mathematics)
Director, Mathematics PreK–12
Baltimore County Public Schools, Maryland

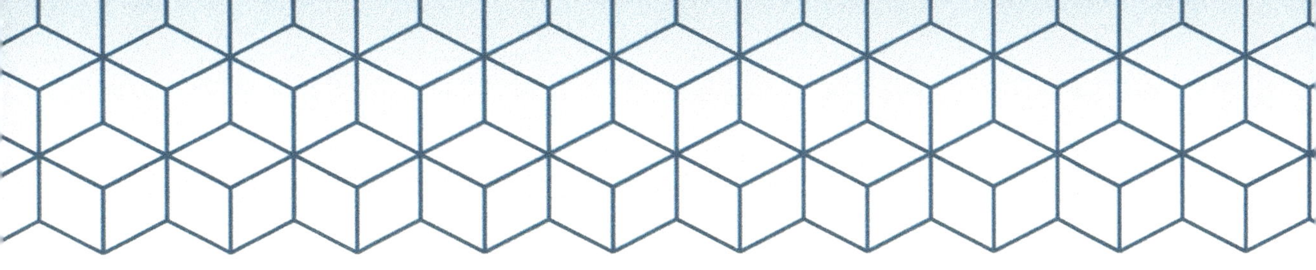

About the Authors

Dr. Fran Arbaugh is an associate professor of mathematics education at Penn State University, having begun her career as a university mathematics teacher educator at the University of Missouri. She is a former high school mathematics teacher, and she received a M.Ed. in Secondary Mathematics Education from Virginia Commonwealth University and a Ph.D. in Curriculum & Instruction (Mathematics Education) from Indiana University–Bloomington. Her scholarship is in the area of professional learning opportunities for mathematics teachers and mathematics teacher educators, and her work is widely published for both research and practitioner audiences. She is a Past-President of the Association of Mathematics Teacher Educators (ATME) and served as a Co-Editor of the *Journal of Teacher Education*.

Margaret (Peg) Smith is a Professor Emerita at the University of Pittsburgh. Over the past two decades, she has been developing research-based materials for use in the professional development of mathematics teachers. She has authored or coauthored over 90 books, edited books or monographs, book chapters, and peer-reviewed articles, including the best seller *Five Practices for Orchestrating Productive Discussions* (co-authored with Mary Kay Stein). She was a member of the writing team for *Principles to Actions: Ensuring Mathematical Success for All,* and she is a co-author of two new books (*Taking Action: Implementing Effective Mathematics Teaching Practices Grades 6–8 & 9–12*) that provide further explication of the teaching practices first described in *Principles to Actions.* She was a member of the Board of Directors of the Association of Mathematics Teacher Educators (2001–2003; 2003–2005), the National Council of Teachers of Mathematics (2006–2009), and the Teachers Development Group (2009–2017).

Justin Boyle is an assistant professor at the University of Alabama. He is interested in learning how best to develop secondary mathematics teachers so that they are prepared to engage their future students in becoming intellectually curious about mathematics. In particular, he uses reasoning-and-proving as a way to investigate and discuss the truth of mathematical statements, concepts, and objects.

Gabriel J. Stylianides is Professor of Mathematics Education at the University of Oxford (UK) and Fellow of Oxford's Worcester College. A Fulbright scholar, he received MSc degrees in mathematics and mathematics education, and then his PhD in mathematics education, at the University of Michigan. He has conducted extensive research in the area of reasoning-and-proving at all levels of education, including teacher education and professional development. He was an Editor of *Research in Mathematics Education* and is currently an Editorial Board member of the *Elementary School Journal* and the *International Journal of Educational Research*. He received an American Educational Research Association Publication Award for his 2009 article "Reasoning-and-Proving in Mathematics Textbooks."

Michael Steele is a Professor of Mathematics Education and Chair of the Department of Curriculum and Instruction in the School of Education at the University of Wisconsin–Milwaukee. He is currently the President-Elect of the Association of Mathematics Teacher Educators. A former middle and high school mathematics and science teacher, He has worked with preservice secondary mathematics teachers, practicing teachers, administrators, and doctoral students across the country for the past two decades. He has published several books and journal articles focused on supporting mathematics teachers in enacting research-based effective mathematics teaching practices. He is the co-author of NCTM's *Taking Action: Implementing Effective Mathematics Teaching Practices in Grades 6–8* and *Mathematics Discourse in Secondary Classrooms*, two research-based professional development resources for secondary mathematics teachers. He is also the author of *A Quiet Revolution: One District's Story of Radical Curricular Change in Mathematics*, a resource focused on reforming high school mathematics teaching and learning.

Setting the Stage

Proof has historically been considered a topic first encountered in high school geometry class, featuring a two-column format of "statements" and "reasons." In this chapter, we present reasoning-and-proving as a set of activities that transcends content and format and is accessible to all middle and high school students. Thinking about reasoning-and-proving more broadly will help you to enhance students' understanding of the mathematics they are learning and their ability to construct valid mathematical arguments. While reading this chapter, we encourage you to consider

- how this broader perspective on reasoning-and-proving could benefit students' understandings of mathematics; and
- what it would take to support students in building the capacity to engage in reasoning-and-proving.

IS REASONING-AND-PROVING REALLY WHAT YOU THINK?

Over the past few years, the mathematics teachers at Hoover High School have been concerned about their students' struggles to think and reason mathematically, which was made salient recently when they reviewed the results of an assessment that featured constructed response items. In general, they found that their students had difficulty explaining why an answer was correct beyond providing a procedural description (i.e., describing *what* they did). The teachers had also noticed that while the students were completing the assessment, they seemed to become quickly frustrated when faced with a task they could not easily and quickly solve. Many of the students' responses were incomplete, and it looked like these students had just given up.

In an effort to improve this situation, all of the teachers in the math department committed to try to engage students in more tasks and activities that emphasize reasoning, justifying, and proving. While the students worked on these problems in small groups, the teachers also worked hard to ask more questions, listen to what students were saying, and to not do so much telling. It has not been easy!

Recently, the algebra teachers have been working on improving their students' abilities to write *proofs*—not the formal two-column variety usually found in geometry—but rather algebraic, visual, and narrative arguments that can be used to explain "why things work" and to verify that something is true and it will work for all cases. In their latest professional learning community (PLC) meeting, they decided to give students the Sum of Three Consecutive Integers task (Figure 1.1). Their students had been working on exploring number theory tasks like this, so the teachers thought this would be a next task for them to do.

Proofs: algebraic, visual, and narrative arguments that explain "why things work" and are true for all cases.

FIGURE 1.1 The Sum of Three Consecutive Integers task.

Prove whether the following statement is true or false: *The sum of any three consecutive numbers is divisible by 3.*

Teaching Takeaway:

Collaborating with colleagues face-to-face or virtually provides an opportunity to share ideas, get feedback on your work, and, in general, support your efforts to improve instruction.

During the PLC meeting, teachers also agreed to identify things that happen during the lesson that they thought were "interesting" and to try to capture these events in some way (e.g., taking notes on a puzzling strategy, collecting samples of interesting student work, recording exchanges with students that they were troubled by). They also agreed to write short vignettes from these artifacts to share with each other. The teachers planned to discuss and analyze the vignettes below at their next PLC meeting.

Vignette 1: Carly Epson's Algebra Class

When I approached Shonda's desk, I noticed that she had created a set of examples that supported the conjecture. She had even written "All of these sums are divisible by three because the sum of their digits is divisible by 3. So it is true because I can't find one that doesn't work."

$2 + 3 + 4 = 9$	[9 is divisible by 3]
$25 + 26 + 27 = 78$	[$7 + 8 = 15$, which is divisible by 3]
$151 + 152 + 153 = 456$	[$4 + 5 + 6 = 15$, which is divisible by 3]
$2744 + 2745 + 2746 = 8235$	[$8 + 2 + 3 + 5 = 18$, which is divisible by 3]

I knew immediately that Shonda's string of examples didn't prove the conjecture, but I didn't want to say that. My hope was by asking the question "How do you know it will always work?" she would have to think twice about what she had done and see the limitations in her approach. But as you can see in the following exchange, it didn't have that effect at all!

Ms. Epson: How do you know it will always work?
Shonda: It has so far and I can't find one that doesn't work.
Ms. Epson: But how can you be sure?
Shonda: I am sure.

She is right—she will never find one that doesn't work, so I certainly didn't want to encourage her to try. But not finding a counterexample doesn't mean it will always work. I wasn't sure what to do next to help her move beyond examples. I told her to keep thinking about how she could convince me. But when I checked back later, she had made no progress.

Vignette 2: Jason Steiner's Algebra Class

When I stopped to check in on Keisha, I observed that she had written the following: "The sum of three consecutive numbers is A + B + C. You can't tell if it is divisible by three or not. There's not enough information."

I decided to start asking her some questions that were intended to move her to a more useful way of representing the three numbers. Here is the gist of the exchange:

Mr. Steiner: So, do you think the statement is true or false?
Keisha: You can't tell because you don't have enough information.

continued>>

<<continued

Mr. Steiner:	What information do you need?
Keisha:	You need to know what one of the numbers is so you can tell what the others will be.
Mr. Steiner:	But suppose that the first number is x. What would the next one be?
Keisha:	y?
Mr. Steiner:	How much bigger than x is the one that comes after x?
Keisha:	1 more.
Mr. Steiner:	1 more. So could you write it as $x + 1$?
Keisha:	I guess so.
Mr. Steiner:	Then what would the next biggest number be?
Keisha:	$x + 2$?
Mr. Steiner:	So can you add those three numbers together?
Keisha:	What three numbers?

I was getting frustrated with Keisha and had no idea what to do next besides telling her what to do. I looked around and noticed that Charles had started on an algebraic solution that resembled the one that I was trying to help Keisha develop, so I suggested that Keisha and Charles share their solutions and decide which one they liked best. In the end, Keisha had a solution that looked like Charles's, but I wondered what she understood about it.

Vignette 3: Barbara Law's Algebra Class

One pair of students, Michael and Marissa, asked if they could go get some tiles. I had no idea what they wanted tiles for, but I told them to help themselves to one of the bins of square tiles. When I checked in with them later, they had arranged the blocks as shown below.

When I asked what the tiles represented, Michael explained, "If the black square is any number, then adding one square to it would be the next number and adding two squares to it would be the one after that." Marissa continued, "If you add the three black squares together, you get three times the number and that is always divisible by three. Then, if you add the three white squares to that number, you will still get a number divisible by three because if you add three to a multiple of three, you get another multiple of three."

I was caught off guard by this approach. I was looking for something more algebraic, and I wasn't sure if this was correct or if it would really count as a proof. I told them it was "interesting" and suggested they try to use algebra to represent the tiles.

Vignette 4: Lynn Baker's Algebra Class

I collected the work from students at the end of class and was not surprised to see that all of the students had clearly defined their variables (x = first number; $x + 1$ = second number; $x + 2$ = third number) because we did this as a class before they wrote their proofs. I was pleased that the students had gone on to show that $x + x + 1 + x + 2 = 3x + 3$. All of the students went on to say that $3x + 3$ had to be divisible by 3 because, as Masey described when she was presenting her group's work, "If you factor out a three, then the threes cancel."

I was really pleased that they all were able to solve the problem and that they got the right answer. Clearly, my students are beginning to understand how to use algebra to prove things!

The work that these algebra teachers are engaged in to improve their students' abilities to think deeply about mathematical concepts and relationships, reason mathematically, and write valid mathematical arguments is commendable. They came together in a PLC to work on their teaching and developed a common goal for improvement that was based on analyzing student data. They chose the same task to implement in their classrooms and gathered artifacts about interesting things that happened during class. They each came away from their class with questions to discuss with each other.

Carly Epson wondered what to do with a student who was convinced that a pattern would always hold true after only testing a few examples. Carly knew that testing examples was not sufficient for arguing the truth of a conjecture, but was unsure what to do next to support Shonda to move to a more generalized argument. Jason Steiner had a similar frustration with Keisha. Barbara Law's student used a pictorial representation to make a generalized argument, and while it seemed convincing, Barbara wondered if this really counted as a proof of the conjecture. And while Lynn Baker felt that her students were well on their way to being able to write algebraic proofs, reading her vignette may have raised questions for you about what her students really knew and were able to do on their own, without her guidance with setting up the variables at the beginning of class.

Pause and Consider

What questions do these vignettes raise for you about mathematical reasoning-and-proving in the context of middle and high school mathematics?

SUPPORTING BACKGROUND AND CONTENTS OF THIS BOOK

We designed this book to support you as you grapple with how to create learning environments that support middle and high school students to become deeper mathematical thinkers, better reasoners, and more capable of making sound mathematical arguments. We believe that this focus is the central work of mathematics education today and that students benefit when learning mathematics in these kinds of environments. We are not alone in this belief.

The field of mathematics education has long emphasized the importance of reasoning and sense-making in K–12 mathematics for the learning of mathematics with understanding. In 1989, the National Council of Teachers of Mathematics (NCTM) drew our attention to this important aspect of learning mathematics by including *mathematics as reasoning* as one of the standards in its seminal publication of *Curriculum & Evaluation Standards for School Mathematics* (NCTM, 1989). NCTM continued to draw our attention to this focus, and began to emphasize the role of proof by including *reasoning and proving* as an essential component of learning mathematics in *Principles and Standards for School Mathematics* (2000) (see Figure 1.2).

A perusal through NCTM's teacher journals (*Teaching Children Mathematics, Mathematics Teaching in the Middle School,* and *Mathematics Teacher*) as well as publications of a number of books over the past 10 years shows that reasoning-and-proving remains in the forefront of the Council's efforts to improve mathematics education in the United States and around the world. Most notably, NCTM (2009) published *Focus in High School Mathematics: Reasoning and Sense Making*. This document, which broadly defined reasoning as ranging from informal explanations and justifications

Instructional programs from prekindergarten through grade 12 should enable all students to—

- recognize reasoning and proof as fundamental aspects of mathematics;
- make and investigate mathematical conjectures;
- develop and evaluate mathematical arguments and proofs;
- select and use various types of reasoning and methods of proof. (p. 56)

Source: NCTM, 2000

to formal deduction or proof, argued "reasoning and sense making should occur in every mathematics classroom every day" (p. 5).

Other standards documents have also advocated for the importance of reasoning-and-proving, with the most recent being the *Common Core State Standards—Mathematics* (CCSSM) (National Governors Association & Council of Chief State School Officers, 2010). CCSSM contains eight *Standards for Mathematical Practice (SMPs)*, which are intended to guide the types of experiences that students need to have while learning the mathematical content standards outlined in the remainder of the document (see Figure 1.3).

Standards for Mathematical Practice (SMPs): eight CCSSM standards that are intended to guide the types of experiences that students need to have while learning the mathematical content outlined in the remainder of the document.

FIGURE 1.3 CCSSM's Standards for Mathematical Practice.

1. Make sense of problems and persevere in solving them.
2. **Reason abstractly and quantitatively.**
3. **Construct viable arguments and critique the reasoning of others.**
4. Model with mathematics.
5. Use appropriate tools strategically.
6. Attend to precision.
7. **Look for and make use of structure.**
8. **Look for and express regularity in repeated reasoning.**

The bolded SMPs are those that connect directly to students' abilities to reason mathematically and engage in mathematical argumentation, although all of the SMPs are supported in learning environments where mathematical reasoning, sense-making, and argumentation are central to everyday work.

How do these kinds of environments benefit middle and high school students? One compelling argument is that these kinds of learning environments support the development of a habit of mind that is useful in mathematics and beyond. In *Focus in High School Mathematics: Reasoning and Sense Making* (NCTM, 2009), the authors argue that reasoning and sense making (which includes making valid mathematical arguments) will "enhance students' development of both content and process knowledge they need to be successful in their continued study of mathematics and in their lives" (p. 7). In particular, reasoning and sense-making skills support informed decision making, promote quantitative literacy, support civic engagement, and position graduates to lead in an increasingly technological economy and workforce (American Diploma Project, 2004; NCTM, 2009). Furthermore, Knuth (2002a) argued that proof (as a form of mathematical justification) can be a powerful tool for the learning of mathematics. Along with actions of engaging in pattern seeking, conjecturing, and writing explanatory proofs, presenting and discussing those proofs publicly benefits students because it "can help demonstrate relationships among areas of mathematics that, to many students, seem unconnected" (p. 489).

By working through this book, you will have the opportunity to more deeply consider the benefits of engaging middle and high school students in reasoning-and-proving through two types of activities:

1. Analyses of narrative cases that feature middle and high school teachers and their students engaged in reasoning-and-proving activities; and

2. Through your implementation of, and reflection on, reasoning-and-proving tasks in your own classrooms.

WHAT IS REASONING-AND-PROVING IN MIDDLE AND HIGH SCHOOL MATHEMATICS?

What does it mean to reason-and-prove in middle and high school mathematics? NCTM (2000) and the CCSSM SMPs (Figures 1.2 and 1.3) provide some guidance. According to the Merriam-Webster

Dictionary (www.merriam-webster.com), *to reason* means to "think in a logical way" and *to prove* means to "establish the existence, truth, or validity of (as by evidence or logic)." We could rewrite these definitions to represent what we mean by reasoning-and-proving in a mathematical context. Instead of defining these terms anew for this book, and in an effort to consolidate the ideas presented in Figures 1.2 and 1.3, we have adopted the phrase *reasoning-and-proving* (Stylianides, 2008b; 2010) to encapsulate this kind of mathematical thinking and argumentation.

Reasoning-and-proving describes the following set of activities: identifying patterns, making conjectures, and providing arguments that may or may not qualify as proofs (Stylianides, 2008b; 2010). In mathematics, the development and validation of new knowledge often passes through several stages, and providing a *proof* is typically the last stage. Earlier stages of mathematicians' work frequently involve exploration of mathematical phenomena to identify and arrange significant observations into meaningful *patterns*. Mathematicians then use those patterns to make *conjectures* and ultimately seek to understand and provide *arguments* about whether and why things work. This progression is represented by the arrows between the stages in Figure 1.4.

This description may infer that the progression through the three stages is linear in nature. In actuality, mathematicians often cycle back to previous stages as they come to understand relationships more deeply, develop counterexamples (which disprove a conjecture), and/or realize that they have not done enough work in the previous stage to move forward. This cyclic progression is represented by the left-facing arrows in Figure 1.4.

Knowing the stages of reasoning-and-proving supports teachers whose students have difficulty providing valid mathematical arguments for a

To Reason: to think in a logical way;

To Prove: to establish the existence, truth, or validity of (www.merriam-webster.com).

Reasoning-and-Proving: the set of activities that includes identifying patterns, making conjectures, and providing arguments that may or may not qualify as proofs.

FIGURE 1.4 Three stages of reasoning-and-proving.

identifying a pattern → making a conjecture → providing arguments (which may or may not qualify as proofs)

conjecture. Let's return for a moment to the Hoover High School teachers' vignettes presented at the beginning of this chapter. Shonda and Keisha had a difficult time providing a valid argument for the conjecture that was already stated in the problem. Would they have been more successful if the conjecture about the sum of three consecutive numbers had not been already given to them? We wonder what would have happened if the task instead had opened up the opportunity to explore and look for patterns—if it had read something like, "Explore the sums of three consecutive integers. What patterns do you see? Make a conjecture and then provide an argument that shows that your conjecture is always true." We do not know for sure what would have happened with Shonda and Keisha, but we would hope that an exploration like this would reveal more about the mathematical structure of the sum of three consecutive numbers than was allowed with the task given.

This three-stage conceptualization of reasoning-and-proving transcends mathematical domains (e.g., algebra, geometry) and representational forms (e.g., algebraic, pictorial) and is based in the work of mathematicians as they seek to develop and validate new knowledge. Although this process of developing new mathematical knowledge may appear linear in nature, it often is more complex. For example, it is not uncommon that when doing this kind of work, mathematicians' efforts to justify a conjecture yield a counterexample as opposed to a proof. This result can prompt further exploration of the original mathematical phenomenon in order to come up with and then justify a new conjecture (e.g., Lakatos, 1976). As we stated previously, reasoning-and-proving is likely to involve movement back-and-forth between the activities of identifying patterns, making conjectures, and providing arguments.

Your work in Chapters 2 and 3 will broaden and deepen your understandings of, and abilities to engage in, reasoning-and-proving.

REALIZING THE VISION OF REASONING-AND-PROVING IN MIDDLE AND HIGH SCHOOL MATHEMATICS

At the heart of this book is our desire to support teachers in realizing the vision of reasoning-and-proving playing a central role in learning mathematics in middle and high school. One of the challenges of realizing the vision of reasoning-and-proving set forth in the standards documents referenced earlier is that many mathematics teachers have not had the opportunity to engage in reasoning-and-proving activities as mathematics learners or to consider how to incorporate reasoning-and-proving into their mathematics teaching. Thus, the purpose of this book is to provide opportunities for mathematics teachers, and those studying to be mathematics teachers, to enhance their own understandings of, and instructional practices for, promoting reasoning-and-proving in mathematics classrooms in Grades 6–12.

Consider again the vignettes presented at the beginning of this chapter. Even though the Hoover High School teachers had good intentions for engaging their students in reasoning-and-proving activities, those intentions were not fully realized. Carly did not know how to move Shonda beyond thinking only through numerical examples, and Jason did not know how to support Keisha to think more abstractly. Barbara was impressed with her students' reasoning, but she wasn't sure that what they had done counted as a proof. Lynn was pleased that her students were "getting it," but as a reader, you may have questions about whether she gave away too much when she set up the task by guiding her students through creating expressions for three consecutive numbers. These teachers had a need to learn new ways of thinking about reasoning-and-proving as well as to develop teaching strategies to support their students' engagement in reasoning-and-proving.

Another challenge of realizing this vision of reasoning-and-proving set forth in these standards documents is that the development of proofs has often been treated as a formal process in geometry and in isolation from the other activities. (You might find it interesting to read about the place of proof in the American school mathematics curriculum in the past century, which you can find in G. Stylianides [2008a]). This treatment of proof has been problematic, because it has not afforded students the level of scaffolding that mathematicians are able to use when making sense of mathematics. Knuth (2002a) wrote about the challenge of meeting the vision set forth by these standards documents and advocated that "adopting a view of proof as a tool for meaningfully learning mathematics, a view underlying the recognitions of the explanatory potential of proofs, gives ways to meet this challenge" (p. 490). This book is intended to help teachers rise to the challenge and put the vision of these standards documents into practice.

In order for students to have increased opportunities to engage in reasoning-and-proving activities, classrooms must be transformed so that understanding and justifying why things work as they do become commonplace. At its heart, reasoning-and-proving involves searching a mathematical phenomenon for patterns, making conjectures about those patterns, and providing arguments demonstrating the viability of the conjecture. In addition, learning mathematics through engaging in reasoning-and-proving requires participation in a community of learners where making thinking public, justifying conclusions, and debating with peers are all hallmark practices of mathematicians. These practices cannot be learned in classrooms where teachers demonstrate how to do procedures and students practice applying learned procedures with no emphasis on sense-making.

Principles to Actions: Ensuring Mathematical Success for All (NCTM, 2014) contains eight effective teaching practices that research indicates support students' learning of mathematics in a deep and connected way. In this book, we use the lens of the effective teaching practices articulated in *Principles to Action: Ensuring Mathematical Success for All,* with a particular emphasis on five of the practices:

1. Establish mathematical goals to focus learning,

2. Implement tasks that promote reasoning and problem solving,

3. Use and connect mathematical representations,

4. Pose purposeful questions, and

5. Facilitate meaningful discourse.

Your work in Chapters 4–6 will focus on considering these five teaching practices through a reasoning-and-proving lens.

MOVING FORWARD

Striving to enhance instructional practices and create learning environments where students are doing this kind of mathematical work is challenging and takes time and patience. If you are reading this book, you have already taken the first step in seeking ways to enhance your own teaching practices and, thus, your students' understandings of important mathematics. You might be in the beginning stages of learning to teach mathematics, perhaps enrolled in a teacher education program at a college or university. You might be a practicing teacher, seeking professional learning opportunities that will support new ways to think about teaching and learning of mathematics. You may be seeking information and advice specifically about supporting your students' abilities to engage in the kinds of thinking, reasoning, and justification in a similar way to the teachers at Hoover High School.

Regardless of where you are in your teaching career, the activities in this book are designed to help you seriously consider the role of reasoning-and-proving in teaching secondary mathematics and the ways in which you can build your students' capacity to engage in these processes. The teachers at Hoover High School, introduced at the beginning of this chapter, identified the need to engage students in more activities that would strengthen their reasoning skills. Despite their best intentions, however, these teachers struggled at times to figure out how to help students make progress on the task without telling them what to do and how. By the time you finish reading this book, we hope that you have new insights into what the Hoover teachers were trying to do and suggestions for them regarding how they might respond to the dilemmas that surfaced during their lessons. Our goal is to equip you with the determination to support your students as reasoners-and-provers and the pedagogical toolkit that will make reasoning-and-proving a reality in your classroom.

This book was written for readers to actively engage while learning more about reasoning-and-proving. There are mathematical tasks to do, student work to analyze, and narrative cases to examine. You will get the most out of the book if you **stop and do** the activities as you progress. Toward that end, we encourage you to keep a journal (separate notebook) as you work your way through the book. Although we have included 10 blank note-taking pages at the end of the book, they are meant for on-the-spot note taking rather than being sufficient for all of your work as you progress through the book. In your journal, you can record your solutions to the mathematical tasks you are asked to do and answers to specific questions raised in the activities, and you can respond to the "Pause and Consider" prompts, which are intended to help you reflect on your learning and consolidate your current thinking at a particular moment in time. Your journal is intended to serve as your personal resource during your journey through the book and as you begin to implement reasoning-and-proving in your own classroom.

Discussion Questions

1. The teachers at Hoover High decided to give their students more opportunities to engage in tasks that focused on reasoning, justifying, and proving. Do all middle and high school students really need to be able to engage in these processes? Why or why not?

2. What opportunities do *your* students currently have to engage in the range of activities associated with reasoning-and-proving (i.e., identifying patterns, making conjectures, providing arguments that may or may not qualify as proofs)?

3. What benefits and challenges would be associated with increasing your students' opportunities to engage in reasoning-and-proving?

Convincing Students Why Proof Matters

Specific examples can be very useful in determining the validity of a mathematical statement or conjecture. In fact, while trying to prove Fermat's Last Theorem (that no three positive integers a, b, and c satisfy the equation $a^n + b^n = c^n$ for any integer value of n greater than 2), many mathematicians started with specific examples. Examples, however, have limitations. In this chapter, you will explore the affordances and limitations of using examples in constructing a proof. In particular, in this chapter you will be encouraged to consider

- how examining specific examples can provide insight to a mathematical statement, but an argument must go beyond specific cases to be valid; and
- how a teacher's actions during instruction can support or inhibit students' ability to reach a mathematical goal such as constructing a valid argument.

WHY DO WE NEED TO LEARN HOW TO PROVE?

Students in middle and high school often question the need for learning how to prove mathematical statements. They see mathematics as a static set of already established rules and procedures that they must learn in order to progress to the next grade. Convincing them that there is a need for proof in mathematics is a challenging task, but one that is well worth the effort to undertake. The activities in the chapter are intended to respond to the perpetual student question, "Why do we need to learn how to prove?"

This chapter draws on work conducted around a sequence of three tasks, which was developed by Andreas Stylianides and Gabriel Stylianides in a 4-year design experiment in an undergraduate mathematics course for prospective teachers. Slightly modified versions of the sequence were implemented both with prospective teachers (G. Stylianides & A. Stylianides, 2009) and high school students (A. Stylianides, 2009) in the United States and the United Kingdom, respectively, and with similarly promising results: A specific and purposeful implementation of the sequence by the instructor in the classroom helped preservice teachers and high school students begin to understand the limitations of empirical arguments and see a need for proof. The careful design of the three-task sequence and its appropriateness for both preservice and high school students contributed to the similar results in the different settings where the sequence was implemented (teacher education program and high school; American and British contexts).

An *empirical argument* is one that uses several examples of specific cases without ever moving beyond the specific to the general. Consider, for example, Shonda's response for proving that the sum of three consecutive integers is divisible by 3, shown in Figure 2.1. Shonda, a student in Carly Epson's class (see Chapter 1), was sure that the sum of three consecutive integers would always be divisible by three because she had not found one example that didn't work.

FIGURE 2.1 Shonda's solution to the Sum of Three Consecutive Integers task (Figure 1.1).

$2 + 3 + 4 = 9$	[9 is divisible by 3]
$25 + 26 + 27 = 78$	[$7 + 8 = 15$, which is divisible by 3]
$151 + 152 + 153 = 456$	[$4 + 5 + 6 = 15$, which is divisible by 3]
$2744 + 2745 + 2746 = 8235$	[$8 + 2 + 3 + 5 = 18$, which is divisible by 3]

This is an empirical argument because it depends on specific examples *and* it does not extend past the specific to a more general argument. Even though Shonda's method falls short of proof, she shared what she was thinking, which can provide her with an opportunity to make progress toward proving if the teacher knew how to support students in the transition from example-based to valid arguments— something that you will learn from this book. As you work on the tasks in this chapter, we encourage you (and for you to encourage your students) to start with what you are thinking even if you believe that what you initially produce is not a proof. We will delve deeper into a discussion about empirical arguments in Chapter 3—for now, this is sufficient for you to engage in the activities in this chapter.

THE THREE-TASK SEQUENCE

This chapter follows, with some adaptations, the version of the task sequence that was produced by A. Stylianides and G. Stylianides and was implemented with high school students by a teacher in a study with A. Stylianides (2009) (available online at http://nrich.maths. org/6664. The sequence includes two types of activities. In Activities 2.1 and 2.2, you will solve a version of two tasks that we adapted from A. Stylianides' (2009) work and then reflect on a third task discussed in the same article. Then you will consider what these three tasks tell us about the limitations of empirical arguments. As you work through the two tasks and consider the third, you will engage in the full range of reasoning-and-proving activities—identifying a pattern, making a conjecture, and providing arguments. In Activity 2.3, you will read and compare the enactment of the sequence of these three tasks in two different mathematics classrooms.

ENGAGING IN THE THREE-TASK SEQUENCE, PART 1: THE SQUARES PROBLEM

The first task in the sequence, Activity 2.1, is the Squares Problem. We encourage you to actively engage with the problem before reading the debriefing text. You can solve the problem any way you like using any materials or representations that you find useful.

Activity 2.1 The Squares Problem

Solve this task:

1. How many different 3 × 3 squares are there in the 4 × 4 square below?

2. How many different 3 × 3 squares are there in a 5 × 5 square?

3. How many different 3 × 3 squares are there in a 6 × 6 square?

4. How many different 3 × 3 squares are there in a 7 × 7 square?

5. How many different 3 × 3 squares are there in a 60 × 60 square?

Source: Adapted from A. Stylianides (2009).

Debriefing Activity 2.1: The Squares Problem

There are many different ways to determine the numeric answers to the questions posed in Activity 1.1. Here we will consider one way to systematically count the number of 3 × 3 squares in the larger squares. Take a few minutes to read through the solutions and, if possible, discuss them with your colleagues. Think about these questions:

- Did your methods yield the same result?
- If not, which squares might you have missed?
- How might you arrive at the totals of the larger squares *without* counting?

Question 1: How many 3 × 3 squares are there in a 4 × 4 square?
One option for counting is to begin counting at the lower left-hand corner of the 4 × 4 square and "sliding" a 3 × 3 square horizontally

   WE REASON & WE PROVE FOR ALL MATHEMATICS

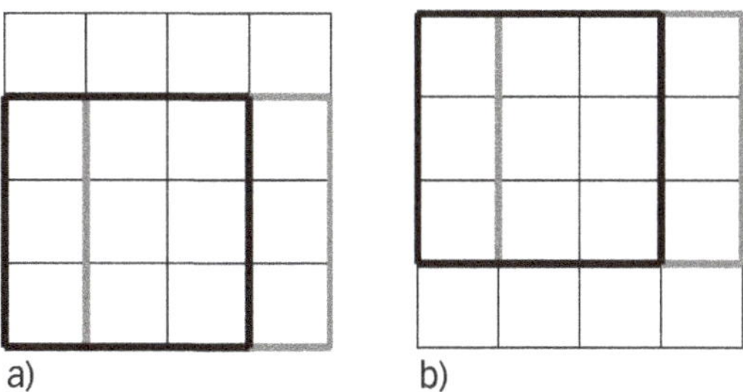

one column to the right, which shows that two 3 × 3 squares will fit horizontally (Figure 2.2a) in the lower portion of the 4 × 4 square. Then we return to the lower left-hand corner and slide the 3 × 3 square vertically one row, and then horizontally one column to show that two more 3 × 3 squares will fit in the upper portion of the 4 × 4 square (Figure 2.2b). So, we can conclude that four 3 × 3 squares fit in a 4 × 4 square.

Question 2. How many 3 × 3 squares are there in a 5 × 5 square?
Using the same counting method as demonstrated in Figure 2.2, we see that there are three 3 × 3 squares that fit horizontally across the bottom portion of the 5 × 5 square (Figure 2.3a), three 3 × 3 squares that fit horizontally in the middle portion of the 5 × 5 square (Figure 2.3b), and three 3 × 3 squares that fit in the top portion of the 5 × 5 square (Figure 2.3c). So, we can conclude that nine 3 × 3 squares fit in a 5 × 5 square.

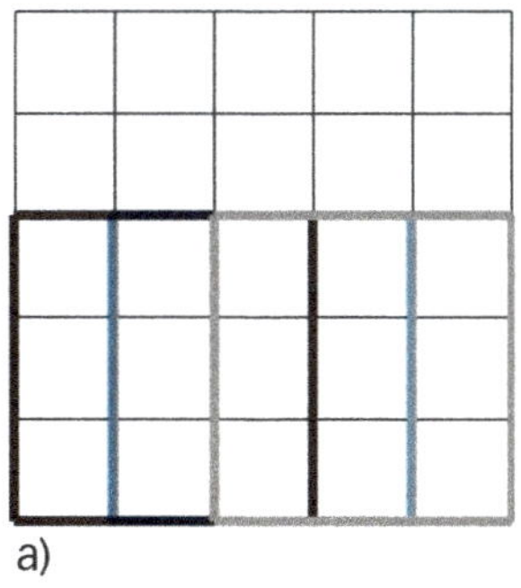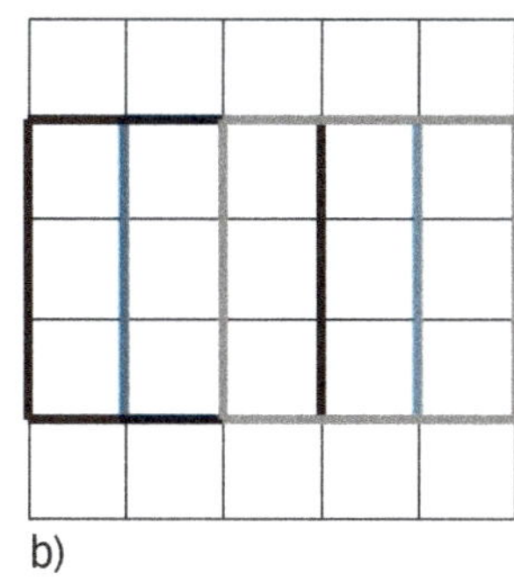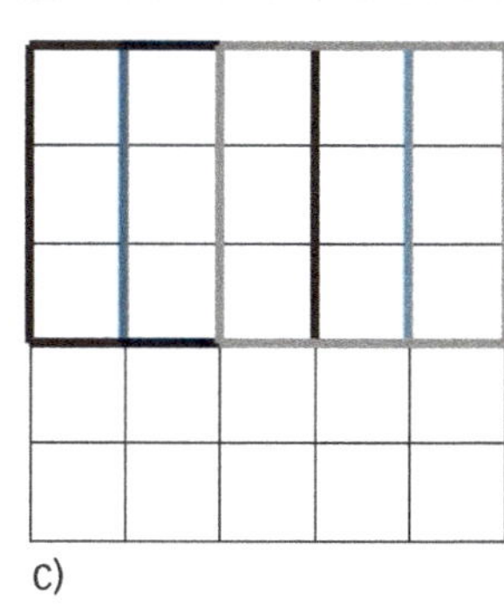

Question 3. How many 3 × 3 squares are there in a 6 × 6 square? Using the same counting method as demonstrated in Figures 2.2 and 2.3, we find that there are four 3 × 3 squares that fit horizontally across the bottom of the 6 × 6 square and that every time we move up vertically one row there are four additional 3 × 3 squares that will fit horizontally. If we continue to move up a row vertically until the 3 × 3 squares will no longer fit, we find that we moved up vertically four times and moved over horizontally four each time. So we can conclude that 16 3 × 3 squares fit in a 6 × 6 square.

Question 4. How many 3 × 3 squares are there in a 7 × 7 square? In the 7 × 7, you will find that you can fit five 3 × 3 squares horizontally and five 3 × 3 squares vertically. This will yield a total of 25 3 × 3 squares in a 7 × 7 square.

Question 5. How many different 3 × 3 squares are there in a 60 × 60 square? Although the same counting strategy used in the smaller cases can be applied here, you may have found that physically counting the squares was more work than you wanted to take on. Instead, you may have looked at the results you found from the first four cases (as shown in Figure 2.4) and noticed a pattern. The number of 3 × 3 squares in a larger square is a perfect square that is two less than the side of the larger square. Applying this to the 60 × 60 square, we would conclude that the number of 3 × 3 squares would be $(60 - 2)^2$ or 58^2. This may have even led you to generalize that the number of 3 × 3 squares in an $n \times n$ square would be $s = (n - 2)^2$.

FIGURE 2.4 The number of 3 × 3 squares that fit in larger squares.

Dimensions of Large Square	Number of 3 × 3 Squares in Larger Square
4 × 4	4
5 × 5	9
6 × 6	16
7 × 7	25

Pause and Consider 

Are you convinced that the number of 3×3 squares in a 60×60 square is 58^2? If you are sure, explain why you are sure. If you are not sure, say why you are not sure. What makes you doubt it? Record your thoughts in your journal or in the blank pages provided at the end of this book.

ENGAGING IN THE THREE-TASK SEQUENCE, PART 2: CIRCLE AND SPOTS PROBLEM

The second task in the sequence, Activity 2.2, is the Circle and Spots Problem. Again, we encourage you to actively engage with the problem before reading the debriefing text. You can solve the problem any way you like, using any materials or representations that would be useful to you.

Activity 2.2 Circle and Spots Problem

Solve this task:

Place different numbers of spots (points) on the circumference of a circle and join each pair of spots by straight lines. Use several drawings of circles and lines to explore a possible relationship between the number of spots and the greatest number of non-overlapping regions into which the circle can be divided. Look on the companion website for a blackline master of pre-drawn circles.

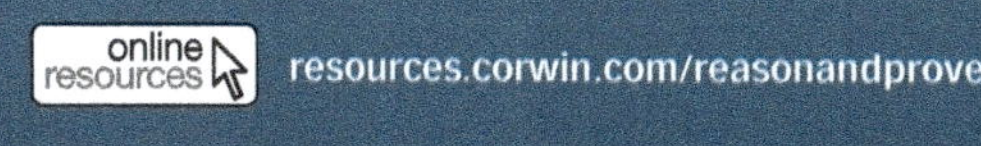

Figure 2.5 contains an example of a circle with four spots.

continued>>

<<continued

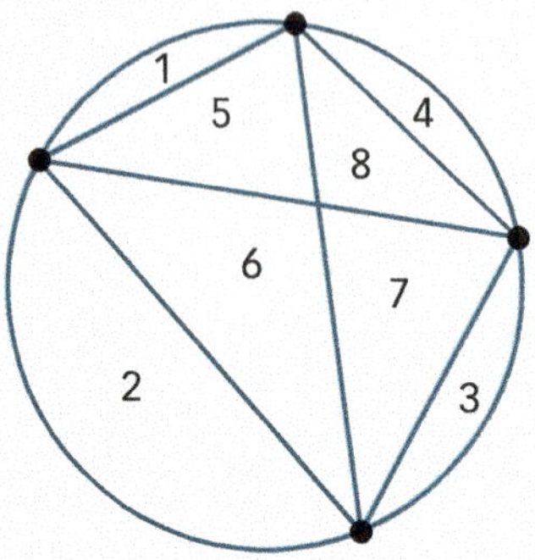

FIGURE 2.5 The number of non-overlapping regions in a circle with four spots on the circumference.

Explore the relationship with other numbers of spots and then answer this question:

When there are 15 spots around the circle, is there an easy way to tell for sure what is the greatest number of non-overlapping regions into which the circle can be divided? What pattern did you notice?

Source: Adapted from A. Stylianides (2009).

Debriefing Activity 2.2: The Circle and Spots Problem

One way to approach this task is to begin by examining smaller cases—that is, consider circles with smaller numbers of spots and determine the maximum number of non-overlapping regions for each, before taking on 15 spots. You may have found that it is pretty challenging to draw 15 spots and actually count the regions!

If you examine the first five cases, and make a list of the number of spots and associated non-overlapping regions (as shown in Figure 2.6), you probably noticed a pattern—the number of regions doubles every time a new spot is added to the circumference. You may have even generalized that the number of regions $r = 2^{n-1}$. Calculating the number of non-overlapping regions when there are 15 spots on the circle then becomes a matter of solving $r = 2^{(15-1)}$.

　　WE REASON & WE PROVE FOR ALL MATHEMATICS

Number of spots	2	3	4	5
Maximum number of non-overlapping regions	2	4	8	16

But what happens with six spots? If you continue the pattern (or used the generalization), you most likely concluded that the number of non-overlapping regions was 32, since this is double the number of regions you got with five spots. However, something very interesting happens when you examine circles that have six spots. Consider the two circles shown in Figure 2.7, each of which shows the number of regions for six spots. The circle on the left has 30 non-overlapping regions while the circle on the right has 31 non-overlapping regions.

How can that happen? At six spots, it starts to matter where you put the spots on the circle as to how many regions are generated—note that the spots on the circle on the left side of Figure 2.7 are relatively evenly spaced around the circle, whereas the spots on the circle on the right side of Figure 2.7 are not evenly spaced. Take a few minutes and examine what happens with the connecting chords (line segments) in the evenly spaced spots and how that is different from the

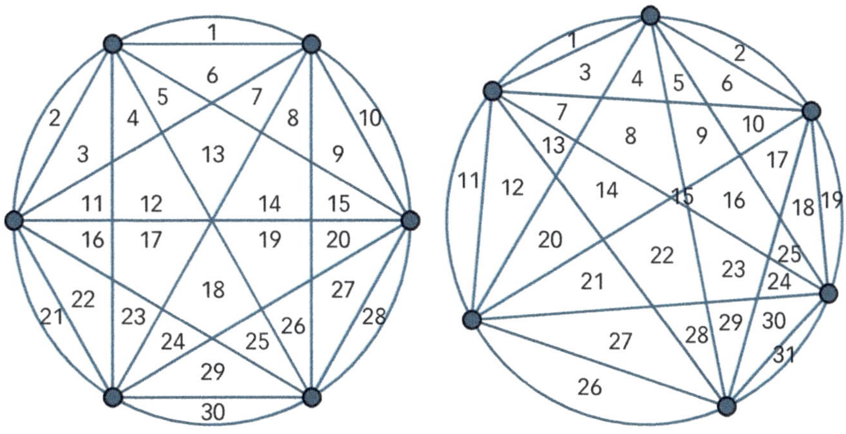

unevenly spaced spots. So, not only do we get two different answers for six spots, the pattern that you may have identified for two to five spots (that the number of regions doubles with the addition of the next spot) no longer holds. No matter where the six spots are placed on the circumference of the circle, the number of non-overlapping regions cannot be more than 31.

Let us now return to the original question that we posed in Activity 2.2: *When there are 15 spots around the circle, **is there an easy way to tell** for sure what is the greatest number of non-overlapping regions into which the circle can be divided?* The answer to the question is simply no, there is not an easy way to tell!

Pause and Consider

The Circle and Spots Problem teaches us that we cannot depend on a pattern to hold true, even though it looks like it will when we start exploring and looking for possible patterns. Why do you think this message is a good one for students to understand when engaged in reasoning-and-proving activities?

One student, when asked what the Circle and Spots Problem taught him, responded as follows:

> *The Circle and Spots Problem teaches me that checking 5 cases is not enough to trust a pattern in a problem. Next time I work with a pattern problem, I'll check 20 cases to be sure.*

Are 20 cases enough to make a strong argument? Some may think so—20 cases that yield a consistent result is pretty convincing evidence. Some may think more cases are necessary. How many cases are needed to make a strong argument that an identified pattern will always hold? Before we answer that question, let's consider one more task and its answer.

   WE REASON & WE PROVE FOR ALL MATHEMATICS

ENGAGING IN THE THREE-TASK SEQUENCE, PART 3: THE MONSTROUS COUNTEREXAMPLE

The third and final task in the sequence is called the Monstrous Counterexample (A. Stylianides, 2009), which we present to you as a story in Figure 2.8.

A group of mathematicians was exploring outputs for the expression $1 + 1141n^2$ when n is a natural number. After evaluating the expression for several natural numbers and looking for any patterns that seemed to be emerging, the mathematicians made the conjecture that this expression *never* returns a square number. To test their conjecture, they wrote a computer program to evaluate the expression for all natural number inputs and to stop when the output tested as a perfect square. Then they left the computer to its work.

They kept checking in with the computer's work, and as n grew bigger and bigger without an output of a square number, they became more and more confident that their conjecture was true. Imagine their surprise when they returned to the computer and found that it had stopped running the program, meaning that the program had an output of a square number. They were even more surprised to find that this expression does return a square number, but not until $n = 30,693,385,322,765,657,197,397,208$.

Source: Stylianides A, J. (2009)

The question we posed before sharing the story of the Monstrous Counterexample was "How many cases are needed to make a strong argument that an identified pattern will always hold?" The story of the Monstrous Counterexample helps to make the point that we can never test enough examples to be assured that an identified pattern does not fall apart at some point. It is interesting to note that if it takes 1 second to test out every n, it would take almost 1 quintillion—a 1 with 18 zeroes—years to test out all of the natural numbers less than the "right" one from the story!

Clearly, testing lots of cases is not productive and can lead to considerable frustration since you will never be sure if you have gone far enough. There has to be a better way! Hence, engaging in this three-task sequence is intended to help students realize that you can never test enough examples to support a strong mathematical claim, and motivate the need to create mathematical arguments that do not depend *solely* on empirical examples. In other words, this three-task sequence helps to motivate a need for proof in mathematics and can be used to convince students why proof matters in mathematics.

Pause and Consider

Think back to the Squares problem that you did at the beginning of this chapter. Were you convinced that you had found the correct answer for a 60 × 60 square? Can you trust the pattern in the Squares Problem based on a few cases when you just saw in the previous two problems that examples cannot be trusted? Record your thoughts in response to these questions in your journal.

We will return to these questions in Chapter 3 after you have engaged in a few more activities.

Empirical Justification Scheme: the misconception that empirical arguments are proofs of mathematical generalizations.

A large body of research over the past few decades has focused on investigating and describing students' knowledge about proof. A robust finding of this body of research is that the *"empirical justification scheme"* (Harel & Sowder, 1998, 2007) is pervasive among school students, including advanced or high-attaining secondary students (e.g., Coe & Ruthven, 1994; Healy & Hoyles, 2000; Knuth, Choppin, Slaughter, & Sutherland, 2002) and university students including mathematics majors (e.g., Goetting, 1995; Sowder & Harel, 2003). In other words, students tend to have the misconception that empirical arguments are proofs of mathematical generalizations. The bottom line is, however, that *empirical arguments* are invalid arguments because they offer inconclusive evidence for the truth of a generalization. They are limited to verifying the truth of only a subset of all possible cases of mathematical relationship suggested in the conjecture. For example, if you think back to Shonda's answer in Figure 2.1, she provided only four valid examples out of an infinite number of possible cases that exist for the relationship of the sum of three consecutive integers. Thus, her response cannot be classified as a proof of the conjecture that the sum of three consecutive integers is divisible by three.

The pervasiveness of this misconception among students is a major stumbling block for the development of their understanding of proof, because when students do not realize the limitations of empirical arguments, they cannot understand why proof is necessary. To date, there has been limited research about how teachers can help students realize the limitations of empirical arguments and see an "intellectual need" (Harel, 2013) for proof. The work of Gabriel and Andreas Stylianides (A. Stylianides, 2009; G. Stylianides & A. Stylianides, 2009) around the three-task sequence, however, indicates that engagement in the sequence can have an impact on how teachers and students think about the limitations of using only examples as the basis for establishing the validity of mathematical arguments.

ANALYZING TEACHING EPISODES OF THE THREE-TASK SEQUENCE: THE CASES OF CHARLIE SANDERS AND GINA BURROWS

In Activity 2.3, you will read and compare the enactment of this sequence of tasks that occurred in two different high school mathematics classrooms. The Cases of Charlie Sanders and Gina Burrows show the task sequence being used with two different groups of students. As you will see, the enactment of this sequence of tasks differs in significant ways in these two classrooms. Our purpose here is not to set up the dichotomy of "bad teaching" versus "good teaching." It is to provide two different classroom enactments of the sequence of tasks and ask you to reflect on the outcomes of the enactments in terms of students' views of reasoning-and-proving as well as on the use of empirical arguments in reasoning-and-proving. Through considering similarities and differences between the two enactments of the task sequence, as well as what Charlie's and Gina's students learned during the enactments, you will begin to articulate reasoning-and-proving *pedagogical moves*, which are actions that teachers take that support (or in some cases, inhibit) students' mathematical engagement and that make a difference to their learning.

Activity 2.3 The Cases of Charlie Sanders and Gina Burrows

In this activity, you will read and analyze two narrative cases of teachers who enact the three-task sequence in their classrooms. You can find *Developing a Need for Proof: The Case of Charlie Sanders* in Appendix A and *Motivating a Need for Proof: The Case of Gina Burrows* in Appendix B.

Read the narrative cases found in Appendices A and B. In your journal, write responses to the following prompts after you have read the two cases.

- What did students in each of these classes learn, and what do you take as evidence of their learning? Cite specific lines in each case to support your claims.
- What are some similarities and differences between the teachers' instruction in these two classrooms?

Debriefing Activity 2.3: The Cases of Charlie Sanders and Gina Burrows

What the students learned. There is little evidence in the Case of Charlie Sanders that students were learning what was intended—*that students start to realize that arguments that are based on the examination of a few examples are not sufficient to prove that something is always true*. Although John's outburst (lines 187–188) suggests that *he* had figured out something important, this was not pursued with John or with anyone else in the classroom. Although Charlie states what should have been learned (lines 231–237), we have no idea whether students actually got this from the lesson.

Throughout the lesson, Charlie failed to draw students' attention to what they should be getting out of each task in the sequence. As a result, students did not have the opportunity to reflect on how each task in the sequence was challenging specific misconceptions they had about empirical arguments being proofs.

Beyond the fact that students' learning of the target mathematical ideas is unclear, it is possible that Charlie's implementation of the sequence was giving his students the following implicit messages about the norms in the classroom:

a) What matters most is the answer to a problem, not the process you used to get there (e.g., lines 96–100, 105–108); and

b) The teacher is the mathematical authority in the classroom (e.g., Charlie recorded only correct answers on the board—lines 98–100; Charlie told groups when they were wrong—lines 105–108).

In comparison, it appears that the ways that Gina implemented essentially the same sequence had a positive impact on her students' understandings about the limitations of using empirical arguments. While students were initially confident in their answer to the number of 3 × 3 squares in a 60 × 60 square based on examining a set of examples, their confidence in examples seemed to waver after their work on the Circle and Spots Problem. While some students seemed to think that more examples would be more convincing, others were not sure. The Monstrous Counterexample story was introduced, not as a task to be completed, but rather as a challenge to the idea that you could never be confident that you have tried enough examples to be sure that something was true. Gina then returned to the Squares Problem and asked students if they could be sure of their answer given it was based on a pattern observed from a small number of cases. The way in which the lesson unfolded suggests that students had an opportunity to learn what was intended—checking examples is not sufficient in making valid mathematical claims and arguments. The exit slip given at the end of class will provide Gina with evidence of what students actually did learn. Using these data, she can then plan her instructional move.

Similarities and differences between the two enactments of the task sequence. While the enactment of the task sequence was very different in these two classes, there were also some things that were very similar, most notably the task and their intentions. In Figure 2.9, we have listed some of the key similarities and associated differences that may have had an impact on students' learning about reasoning-and-proving.

Similarities	Differences
Charlie and Gina had the same overarching goal for instruction—to help students see the limitation of empirical arguments in proving mathematical conjectures. However…	Charlie did not appear to achieve the goal; it is not clear that he fully understood the learning path—what each task was intended to contribute to students' understanding or what role he needed to play to scaffold students' learning. Gina appeared to achieve her goal; she reached a point with her students where they appeared to realize that they couldn't trust examples.
Charlie and Gina used the same sequence of tasks. However…	Although Charlie "did" the tasks, he did not ask questions that would have ultimately helped the students see the big picture. For example, although after initial work on the Squares Problem he asked students if they were sure the formula was true, he did not ask them how they got the formula or why they thought it would hold up. So, the whole idea of only trying a few cases was never really made salient. From the beginning of the sequence, he did not lay a foundation on which to build. Gina, starting with the Squares Problem and, continuing with the Circle and Spots Problem, built a clear case about the role examples played in the conclusions students were drawing, which set them up for the Monstrous Counterexample story and the need for more secure methods of validation.
Charlie and Gina both pressed students to think about whether they were sure about their answers to Part 3 of the Squares Problem. However…	Charlie asked students if their answer to Part 3 was always true, which appeared to baffle students—they seemed to have no idea why he asked this question. Gina made clear that they were only basing their conclusion to Part 3 on the examples they tested.

Three other differences are noteworthy:

- **Interactions with small groups.** Charlie monitored what students were doing and indicated when students were incorrect, but never pressed students to explain what they were doing or asked them any questions to push their thinking. Gina visited small groups and asked students to explain their work. She asked questions such as, "What were you thinking about? How do you think that might help you?"

- **Use of whole-class discussion to expose and solidify students' thinking.** Charlie did nearly all of the talking during whole-class discussion and as a result got very little insight into what students were thinking. Gina conducted a discussion after work on each problem; during the discussions, she asked students to explain things and asked students both whether and why they were sure of their answers.
- **Use of formative assessment.** There was no evidence that Charlie used formative assessment. At three points during Gina's lesson (after the initial work on the Squares Problem, after work on the Circle and Spots Problem, and at the end of the lesson), she asked students to individually write down their thoughts. This provided her with immediate feedback on how students were making sense of the lesson, which she could use as she continued to move forward.

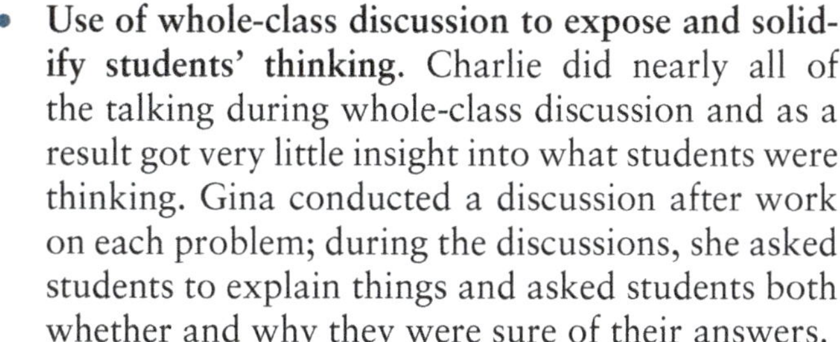

Pause and Consider

In analyzing the Cases of Charlie and Gina, we noted that the implementation of the task sequence was very different in the two classes and that Gina was able to achieve her goal for the lesson while Charlie was not. What did Gina do that made it possible for her to accomplish what Charlie could not?

Start a page in your journal or use one of the blank pages in the back of the book and title the page "Pedagogical Moves That Support Students' Capacities to Reason-and-Prove." Begin to articulate teaching moves/actions that appear to support students' capacities to reason-and-prove. We will revisit and add to this list in other places in the book, so leave room for additions.

CONNECTING to Your CLASSROOM

As we close this chapter, we suggest that you consider implementing the three-task sequence in at least one of your math classes. If you have engaged with all of the activities in this chapter, we believe that you are prepared to implement the sequence with students. To support your implementation, we have developed an implementation guide, which you can find on the book's website.

resources.corwin.com/reasonandprove

Download an implementation guide for the three-task sequence.

Note that in the lesson plan, we suggest a short exploration of the Monstrous Counterexample problem before showing students the answer. You can choose to implement this part of the lesson either as suggested in the Case of Charlie Sanders and the lesson plan, or as we did in this chapter with a story, and as Gina Burrows did. We also suggest that you enlist a colleague or colleagues to implement this sequence too, so that you can plan together and then debrief what happened.

If you and colleagues implement this sequence, then during a debriefing session, discuss what you think students learned, what evidence you have to support your claim, and identify what you did to support (or inhibit) students' learning. Discuss what you will do differently (if anything) the next time you teach this lesson.

*****

In this chapter, you have engaged in activities designed to make you think hard about the limitations of empirical arguments in reasoning-and-proving. We hope that you are now convinced that there is a need for proof—that proof matters in mathematics! You have also begun to generate knowledge about what a teacher can do to support students' capacities to reason-and-prove by identifying pedagogical moves that support this kind of learning environment. Your work in Chapter 3 will extend your understandings of reasoning-and-proving, and in Chapter 4, we will return to thinking about teaching in ways that support your students' capacities to reason-and-prove.

1. In the Hoover High School Case presented in Chapter 1, we saw Shonda (a student in Carly Epson's class) create an empirical argument showing the sum of three consecutive numbers was divisible by 3. She showed that the statement was true for the cases she tried but did not explain *why* it was true. What might a proof of this statement look like? Why is it important to show something is true for all cases?

2. In the Hoover High School Case presented in Chapter 1, we saw several teachers (Carly Epson, Jason Steiner, Barbara Law) struggle to support students in creating arguments that would qualify as proofs. What strategies did you identify in your analysis of the Gina Burrows lesson that might have helped the teachers at Hoover High better support their students in creating valid arguments?

3. What strategies used by Gina Burrows might be generally helpful to you in engaging your students in learning mathematics?

Exploring the Nature of Reasoning- and-Proving

There are many different and valid ways to construct a proof. For example, there are more than 350 different proofs of the Pythagorean Theorem. In order to qualify as a proof, however, certain criteria must be met and the proof must be accepted as valid by one's peers. In 1993, when Andrew Wiles first presented his proof of Fermat's Last Theorem, his peers critiqued it and identified flaws in his reasoning. He then revised his proof and resubmitted it for others to critique before it was accepted as valid in 1994. In this chapter, you will engage in activities that will help you articulate answers to the questions, "What kinds of activities constitute reasoning-and-proving?" and "How can you judge whether an argument is a proof or not?" In particular, in this chapter you will be encouraged to consider

- when a mathematical argument rises to the level of proof;
- the criteria that need to be met in order to classify mathematical arguments as proofs; and
- what constitutes reasoning-and-proving.

WHEN IS AN ARGUMENT A PROOF?

You likely bring a host of ideas about reasoning-and-proving into this chapter. For some readers, engaging with the content of this chapter may confirm pre-existing beliefs and knowledge about reasoning-and-proving. For others, the chapter may challenge the ways that you currently think about reasoning-and-proving. In Chapter 2, we made salient why empirical arguments cannot be considered proofs. In this chapter, you will develop a set of criteria that will help you and your students answer the question, "When does a mathematical justification or argument constitute a proof?" We then present a framework about reasoning-and-proving that serves as a tool for situating "proof" in the broader set of activities that comprise the notion of reasoning-and-proving.

Your work in this chapter begins with completing a proof task based in number theory. While you engage in writing the proof, pay close attention to how you are thinking about what it means to write a proof of a conjecture. Those thoughts will come into play as you engage in subsequent activities in this chapter.

Activity 3.1 Creating a Mathematical Argument

The Sum of Two Odds Task

Prove that when you add any two odd numbers, your answer is always even. Solve the task in any way you want to using any representations or materials that will help you make sense of the situation.

Debriefing Activity 3.1: Creating a Mathematical Argument

There are many ways to prove that the sum of two odd numbers is even. You may have started by looking at specific examples that convinced you that this conjecture was true. The problem with this approach is that there is no way of knowing if there is a case that doesn't work unless you try them all—and you can't try them all! Remember the Circle and Spots Problem in Chapter 2? There you may have been convinced that the number of nonoverlapping regions formed by connecting n spots could be expressed by the generalization $(r) = 2^{n-1}$—until you tried six spots. So we know that just examining cases is not enough.

One way to think about proof is to consider how you could convince a skeptic that the conjecture will always be true. Key to proving this conjecture is recalling what it is you know about odd and even numbers. For example, an even number is divisible by two and an odd

    WE REASON & WE PROVE FOR ALL MATHEMATICS

number has a remainder of 1 when divided by 2—an odd number is one more than an even number. This reasoning could lead you to an algebraic solution. This is a popular strategy among the teachers with whom we have worked. The key to this approach is representing an odd number in terms of an even number ($2x + 1$) and ensuring that you are adding two unique odd numbers (e.g., $2x + 1$ and $2y + 1$). If you used an approach of this type, consider what makes it a proof and how you could represent this in some other way that could be convincing to someone who did not know algebra.

Rather than provide a full set of possible proofs to the conjecture here, we invite you to move on to Activity 3.2, where you will analyze a set of student solutions to this task and consider which of the solutions should be considered a proof. As you read through the solutions, think about which one most closely resembles the one you produced.

Activity 3.2 Analyzing Mathematical Arguments

1. Examine the 10 student responses (labeled A-J) to the "Sum of Two Odds" task in Figure 3.1. Look for the "Sum of Two Odds" task recording sheet on the companion website.

2. For each of the 10 responses, make a judgment as to whether or not the student has proven the conjecture. After you decide (Is it a proof?), write yes or no in the second column of the recording sheet and record an explanation for why you made that decision in the third column. You might be tempted to say that a particular piece of student work is "almost a proof"—if so, you should assign "no" to that piece of student work and write a statement about what would need to be done to make it a proof.

3. After you have judged all of the student responses, look across the explanations that you wrote for justifying why a response did or did not qualify as a proof. Generate a list of statements in your journal that summarizes your explanations. You can start the list with the sentence stem, "To be judged a proof, an argument must ...". You will be using this list in Activity 3.3.

Student Response	Proof? (Yes or No) Explanation
Student A* If I take the numbers 5 and 11 and organize the counters as shown, you can see the pattern. 5 + 11 You can see that when you put the sets together (add the numbers), the two extra blocks will form a pair and the answer is always even. This is because any odd number will have an extra block and the two extra blocks for any set of two odd numbers will always form a pair. 16	
Student B** $3a + 3b = 6(a + b)$ $a = 3$ $b = 9$ $(3 \times 3) + (3 \times 9) = 36$ $5a + 5b = 10(a + b)$ $93a + 57b = 140(a + b)$ An even number of odd numbers make an even answer but an odd number of odd numbers makes an odd answer: Odd + Even $7a + 9b = 16(a + b)$ Odd + Even + Odd $7a + 9b + 11c = 27(a + b + c)$ Odd + Even + Odd + Even $7a + 9b + 11c + 13d = 40(a + b + c + d)$ Odd + Even + Odd + Even + Odd $93a + 7b + 13c + 101d + 39e = 153(a + b + c + d + e)$	

Student Response	Proof? (Yes or No) Explanation
Student C Any odd number can be written as $2x + 1$. So, let's add two odd numbers. $2x + 1 + 2x + 1 = 4x + 2$ $4x + 2$ is even since 4 and 2 are both even. Or $2(2x + 1)$ shows that $4x + 2$ is even.	
Student D** My answer	

add 1 (a)	add 2 (b)	$a + b$
1	3	4
7	9	16
11	13	24
21	23	44
113	97	210
1111	1111	2222
1003	10003	11006

I noticed all the sums will be an even number.
$a + b = c$
Test: $a = 35$, $b = 73$, then $35 + 73 = 108$. 108 is also even so it is true.

continued>>

Student Response	Proof? (Yes or No) Explanation
Student E* If I take the numbers 5 and 11 and organize the counters as shown, you can see the pattern. 5 + 11  You can see that when you put the sets together (add the numbers), the two extra blocks form a pair and the answer is even. 16 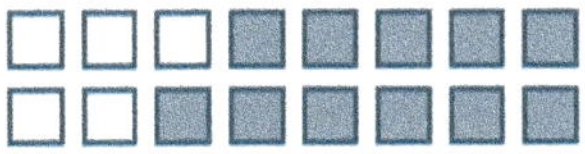	
Student F* An odd number = [an] even number + 1. e.g. 9 = 8 + 1 So, when you add two odd numbers you are adding an even no. + an even no. + 1 + 1. So, you get an even number. This is because it has already been proved that an even number + an even number = an even number. Therefore, as an odd number = an even number + 1, if you add two of them together, you get an even number + 2, which is still an even number.	
Student G* If you add two odd numbers, the two ones left over from the two odd numbers (after circling them by twos) will group together to make an even number.	

<table>
<thead>
<tr><th>Student Response</th><th>Proof? (Yes or No)
Explanation</th></tr>
</thead>
<tbody>
<tr><td>

Student H

Definition of an even number: An integer p is even if and only if there is an integer k such that $p = 2k$

Definition of an odd number: An integer q is odd if and only if there is an integer k such that $q = 2k + 1$. Let's assume X and Y are odd where $X = 2n + 1$ and $Y = 2m + 1$, and n and m are integers.

Statement	Reason
X is an odd number so $X = 2n + 1$	Given, Definition of odd number
Y is an odd number, so $Y = 2m + 1$	Given, Definition of odd number
$X + Y = 2n + 1 + 2m + 1$	Addition Property of Equality
$X + Y = 2n + 2m + 1 + 1$	Commutative Property of Addition
$X + Y = 2n + 2m + 2$	Substitution
$X + Y = 2(n + m + 1)$	Distributive Property
$n + m + 1$ is an integer	Closure Property of Addition for Integers
$X + Y$ is an even number	Definition of an even number

</td><td></td></tr>
</tbody>
</table>

continued>>

Student Response	Proof? (Yes or No) Explanation

Student I

An odd number has to have an odd digit in the ones place. When you add any two single-digit odd numbers you would get an even number in the ones place. Here are all of the numbers you get when you add two single-digit odd numbers.

+	1	3	5	7	9
1	2	4	6	8	10
3	4	6	8	10	12
5	6	8	10	12	14
7	8	10	12	14	16
9	10	12	14	16	18

The ones place is the only place that matters in determining if a number is odd or even so it doesn't matter how many other digits the number has. If it is odd it will always have a 1, 3, 5, 7, or 9 in the ones place. If it is even, it will always have a 0, 2, 4, 6, or 8 in the ones place.

Student J*

If a and b are odd integers, then a and b can be written $a = 2m + 1$ and $b = 2n + 1$, where m and n are other integers.

If $a = 2m + 1$ and $b = 2n + 1$, then $a + b = 2m + 2n + 2$.
If $a + b = 2m + 2n + 2$, then
$a + b = 2(m + n + 1)$.
If $a + b = 2(m + n + 1)$, then
$a + b$ is an even integer, by the definition of an even integer.

* Student responses A, E, G, and J were adapted from Coxford et al. (1999).
** Student responses B, D, and F first appeared in Healy and Hoyles (2000).

    WE REASON & WE PROVE FOR ALL MATHEMATICS

Debriefing Activity 3.2: Analyzing Mathematical Arguments

Engaging in this activity may have challenged you to think hard about what it means to prove a mathematical claim or conjecture. If you worked through this activity with colleagues, we suspect that there were disagreements that had to be negotiated. If you worked through it on your own, then reading this debriefing may challenge your beliefs about what it means to prove. In our work with teachers, engaging in this activity has generated healthy debate on what it means to prove. This activity has also been a powerful learning experience for the teachers—they have been forced to consider new ideas about proof and have had long-standing ideas about what makes a "good proof" challenged. Many of the teachers have never been asked to articulate what makes an argument a proof, making this activity intellectually challenging. So, if your "brain hurts" at the end of this activity (as many teachers have reported), that's okay—it means that you have been thinking deeply about these ideas.

Our analysis of student responses A-J is found in Figure 3.2. Take a few minutes to read the contents of the table, working back and

FIGURE 3.2 Explanations for assigning yes/no to Student Responses A–J.

Student Response	Proof?	Explanation
Student A	Yes	The student begins with a specific example, but expands the argument to cover all odd numbers. The use of a concrete model to represent an odd number assumes that the claim is valid for the set of natural numbers, but this is not explicitly stated. Odd and even numbers are not explicitly defined, which assumes that this is a known fact or has been previously proven.
Student B	No	This student has created an argument that contains many statements that show a misunderstanding of the distributive property. While the student has discovered patterns in the sums of odd numbers by testing specific cases, the student has made no progress in proving the claim. The use of variables here serves no purpose. There is much work that needs to be done to make this a proof, including finding a way to represent or describe an odd number in general terms, as seen in the responses produced by Students A and F.

continued>>

Student Response	Proof?	Explanation
Student C	No	The student created a general argument, but it only deals with the specific case where an odd number is added to itself. In order to make this a proof, the student needs to use two unique odd numbers ($2x + 1$ and $2y + 1$).
Student D	No	The student has constructed an empirical argument that uses a set of specific examples. Although the student uses numbers of different sizes, there is no attempt to generalize beyond the specific cases used. In order to make this a proof, the student would need to move from the specific examples to the more general case. This would begin with finding a way to represent or describe an odd number in general terms, as seen in the responses produced by Students A and F.
Student E	No	The student uses a single example to argue the claim. In order to make this a proof, the student would need to move from the specific example to the more general case, as seen in Student A's solution.
Student F	Yes	The student defines an odd number as an even number + 1 and illustrates this with a numerical example. Even numbers are not defined, which assumes that the definition of even numbers is known. The student uses a previously proven statement (i.e., even + even = even) effectively.
Student G	No	The student is making a general argument about the addition of two odd numbers, but it relies on many things being known to the reader. In order to make this a proof, the student needs to define odd and even numbers and provide more explanation regarding why the process of "circling by twos" makes sense in this situation.
Student H	Yes	The student provided a general argument, defined all terms, and justified every step.
Student I	Yes	The student showed every case of what happens when adding two one-digit odd numbers and explained why it would hold for larger odd numbers. The student assumes that the claim is restricted to the set of natural numbers (since negative values are not considered). The student also utilizes the fact that odd numbers have odd digits in the ones place and even numbers have even digits in the ones place.
Student J	Yes	The student defines an odd number in terms of an even number, which assumes that the definition of even numbers is known or has been previously proven. The student specified the set of numbers to which both m and n belong and provided a logical algebraic argument.

forth between our responses and your responses. Make notes in your journal about agreements and disagreements and whether our explanations led you to reconsider any of your initial judgments. If possible, discuss these proof/not a proof categorizations and explanations with other teachers.

You may have noticed a number of things while you were analyzing the student work and considering our explanations in Figure 3.2. First, you probably recognized Student H's response as the formal two-column proof that is commonly used in geometry and correctly classified it as a proof. You may have also considered Student J's algebraic solution, similar to Student H's, to be a proof. Although Student C also uses algebra, the response shows only that when you add the same odd number to itself you get an even number. Hence it is not general—it does not examine all cases. While Student B also attempts to use algebra, the variables serve no purpose and do not make clear what an odd number is.

Second, you may have noticed that Students A, D, and E all used examples, but that only the response from Student A was considered a proof. The distinguishing feature between these three responses is that Student A actually moves from the example to a more general case (**"any odd number** will have an extra block and the two extra blocks for any set of two odd numbers will always form a pair"). You may not have considered Student I's response to be a proof since it seems to be using specific cases. The critical feature here is that the proof "exhausts" every possibility for the digit that appears in the ones place—the only place value that matters in determining whether a number is odd or even.

Finally, Students F and G both make what we would call narrative arguments, but only Student G has constructed a proof. The difference here is that while Student F makes clear what constitutes an even and an odd number, Student G assumes too much—they need to define odd and even numbers as the basis for the argument.

In Question 3 of Activity 3.2, we asked you to generate statements in your journal about the components of an argument that are necessary in order to be counted as a proof. In Activity 3.3, we ask you to review this list in light of the analysis of student responses in Figure 3.2.

Activity 3.3 Developing Criteria for Determining If an Argument Counts as a Proof

1. In Part 3 of Activity 3.2, you developed an initial list of criteria for judging whether an argument is a proof. After reading our explanations in Figure 3.2, return to your list of statements and for each item on your list, ask yourself, "Does this item on my list really matter in determining whether an argument is a proof?" Make adjustments to your list as needed.

2. Review our list of criteria for judging whether an argument counts as a proof, shown in Figure 3.3. Compare the list in Figure 3.3 to your list of criteria. Are these four criteria on your list? If your list has extra criteria, reflect on why those items may not be on our list.

FIGURE 3.3 List of criteria for judging whether an argument counts as a proof.

Criteria for Judging Whether an Argument Counts as a Proof

An argument that counts as a proof must meet all of the following criteria:

1. **The argument must show that the conjecture or claim is (or is not) true for *all* cases.** *(Student H is an example of an argument about all cases of sums of two odd numbers. While Student C also provides an algebraic argument, it deals only with the special case of adding any two of the same odd numbers, so it does not meet this criterion.)* Specific examples *can* be used in the argument, but it is essential at some point to move from particular examples to a discussion of the more general case. *(While both Students A and E use a specific example, only Student A moves from the specifics of adding 5 + 11 to the more general case of considering any two odd numbers. Therefore, Student A's argument would be considered a proof, but Student E's argument would not.)*

2. **The statements and definitions that are used in the argument must be ones that are true and accepted by the community because they have been previously justified.** *(Student F used the statement that even + even = even in her proof that odd + odd = even. Since she goes on to say that this statement had been previously established, it was appropriate to use.)*

3. **The conclusion that is reached from the set of statements must follow logically from the argument made.** *(Student H's conclusion that X + Y is an even number follows logically from the set of statements made beginning with defining X and Y to represent odd numbers 2n + 1 and 2m + 1 respectively; Student F's conclusion that "if you add two of them together, you get an even number + 2, which is still even" follows from the argument made regarding an odd number being the sum of an even number and 1.)*

4. **The mathematics must be correct.** *(Among other problems with Student B's response, there are mathematical mistakes in this argument. For example, Student B writes that 5a + 5b = 10(a + b), which is incorrect.)*

Debriefing Activity 3.3: Developing Criteria for Proof

There is no exact agreement on proof criteria in the broader mathematical community. However, it is important to note, from a mathematical argument perspective, that this list of criteria *cannot be arbitrary*—it is not the case that "anything goes" when developing this list. Whether an argument constitutes a proof *is not* a matter of opinion. Criteria used to judge whether an argument is a proof must be consistent with ideas about proof that are widely accepted in the broader mathematical community, as the criteria in Figure 3.3 do.

Students often leave secondary geometry classes believing that all proofs must be organized in two columns, with those two columns labeled "statement" and "reason" (Harel & Sowder, 1998). As can be seen from this activity, proofs can be presented in a number of ways. For example, Student F presented a proof in narrative form. Student I also used a narrative form and a table of values to create a proof by exhaustion. Students also leave

Teaching Takeaway:

The list of criteria for judging whether an argument counts as proof *cannot be arbitrary*—it is not the case that "anything goes" when developing this list.

Teaching Takeaway:

Proofs do not have to be organized in a two-column format. They can be presented in a number of ways.

high school believing that all proofs must use deductive reasoning, which can cause some problems when they encounter different argument schemes in college mathematics (e.g., proof by contradiction, proof by induction). Student A presented a proof in which deductive reasoning is not used. We will classify this type of proof later in this chapter.

We began this chapter by suggesting that you likely brought a host of ideas about reasoning-and-proving, as well as what constitutes a proof, to the reading of the chapter. Engaging in Activities 3.2 and 3.3 may have prompted you to reconsider long-held beliefs about what proofs "must" look like or contain, and we hope that by the time you have engaged with the activities in this book as a whole, you can see that the items listed below are not a "must" for assessing the validity of an argument. They are, instead, personal preferences for the presentation of proofs—these items do not impact the validity of a mathematical argument. To summarize, a proof *may* vary along the following dimensions:

- *type of proof* (e.g., direct proof, proof by induction, proof by exhaustion)
- *form of the proof* (e.g., two-column, narrative, flow chart)
- *representation used* (e.g., symbols, pictures, words)
- *explanatory power* (e.g., how well the proof itself serves to explain why the claim is true)

Variance on these dimensions *does not matter* to whether an argument counts as a proof as long as the four criteria presented in Figure 3.3 are met.

You may want to engage your students in developing a list of criteria to use to judge whether an argument is a proof (which many teachers have done with success). The process of developing a list of criteria will help ensure that the community of learners in your classroom are in agreement about what is needed in order to call an argument a proof. We also believe that engaging in this activity will cause students to grapple with notions of what a "proof" is and support their learning in deeper ways than if you simply give them a set of criteria to use when judging whether an argument counts as a proof.

When engaging your students in the process of developing criteria, it is important that you orchestrate the discussion so that the list of criteria your students generate reflects the four key ideas embedded in the criteria shown in Figure 3.3. As teachers of mathematics ourselves, we think that it is vitally important to ensure that the criteria negotiated and ultimately accepted in a classroom community conform with disciplinary standards (A. Stylianides & G. Stylianides, 2009). To support your implementation of Activities 3.1–3.3 with your students, we have included an implementation guide on the companion website along with a printable version of Figure 3.1.

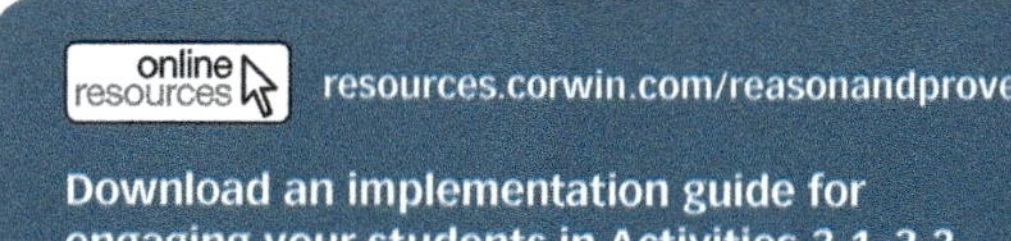

THE REASONING-AND-PROVING ANALYTIC FRAMEWORK

As we explained in Chapter 1, the development of the activities in this book was guided by Gabriel Stylianides' (2008a, 2010) conceptions of reasoning-and-proving, which he defines as including the following set of activities: identifying patterns, making conjectures, and providing arguments that may or may not qualify as proofs. In this section, we expand on Stylianides' definition of reasoning-and-proving by presenting and discussing his Reasoning-and-Proving analytic framework.

Stylianides' analytic framework for reasoning-and-proving (Stylianides, 2008a, 2010) has three components: mathematical, learner, and pedagogical (teaching), as represented in Figure 3.4. Taken together, this framework "integrates several major constituent activities of reasoning-and-proving, and supports teachers to think in connected ways about pertinent mathematical, learning, and pedagogical issues" (2010, p. 44).

Reasoning-and-Proving				
Mathematical Component	What are the major activities involved in reasoning-and-proving?			
	Making generalizations		Developing arguments	
	Identifying a pattern (plausible or definite)	Making a conjecture	Developing a proof (generic argument or demonstration)	Developing a non-proof argument (empirical argument or rationale)
Learner Component	What are students' perceptions of the mathematical nature of a pattern/conjecture/proof/non-proof argument?			
Pedagogical Component	How does the mathematical nature of a pattern/conjecture/proof/non-proof argument compare with students' perceptions of this nature? How can teachers help their students reconsider and change (if necessary) their perceptions to better approximate the mathematical nature of a pattern/conjecture/proof/non-proof argument?			

Source: G. Stylianides (2008a, 2010). Reprinted with permission from *For the Learning of Mathematics.*

The learner component of the framework "focuses on the student and is intended to draw attention to the student's perception of the mathematical nature of a pattern/conjecture/proof/non-proof argument" (Stylianides, 2010, p. 219). Understanding the students' perceptions of reasoning-and-proving can support you in engaging students in reasoning-and-proving activities by providing a window into what students think is important in reasoning-and-proving as well as how they might approach a reasoning-and-proving task. You can then use that information to assess how well students' perceptions cohere with the mathematical notions of reasoning-and-proving, which are captured in the mathematical component of the framework and explained in more depth below.

Then, based on what you hear from students, you make a pedagogical move (e.g., provide a task, ask a question, challenge an assumption) that helps your students develop greater understanding of the

Teaching Takeaway:

The Reasoning-and-Proving Framework can support you in helping your students enhance their capacities to reason-and-prove.

reasoning-and-proving process, thus employing the pedagogical component of the framework. For example, if a group of students perceives that empirical arguments are sufficient for proving a conjecture (like Student D in Figure 3.2), knowing that information can help you engage the students in ways that challenge this common misconception (by enacting, for example, the series of tasks from Chapter 2). If a student believes that symbolic algebra is required for a proof to be valid, then you might introduce a conjecture or task situation that is difficult to prove algebraically, like the Squares Problem in Chapter 2, which can be proven via proof by induction using algebra—but that is pretty sophisticated for most secondary students. At the end of this chapter, we discuss how we *can* prove the conjecture generated from your work on the Squares Problem in a way that is accessible for secondary students.

The mathematical component of the framework is a bit different from the learner and pedagogical components. It has multiple parts and answers the question, "What are the major activities involved in reasoning-and-proving?" In Chapter 1, you read about the stages of the reasoning-and-proving process: identifying a pattern, making a conjecture, then providing an argument (which may or may not qualify as a proof). Recall that working through these stages may not be linear—that the mathematical work done during reasoning-and-proving activities may include revisiting previous stages before completing a mathematical argument. We unpack the three stages further in the sections below. As you learn about and make sense of the mathematical component of the framework, it will be helpful throughout these next sections to continually refer to the Reasoning-and-Proving Framework found in Figure 3.4. Look for a printable version of the framework on the companion website.

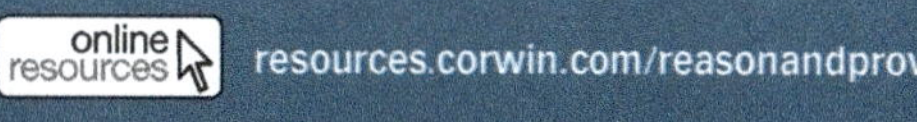

Identifying a Pattern

Children spend a lot of time in elementary school on patterning activities. In early educational settings, they make patterns with colored blocks and counting manipulatives and learn to continue patterns started by others, as represented in Figure 3.5.

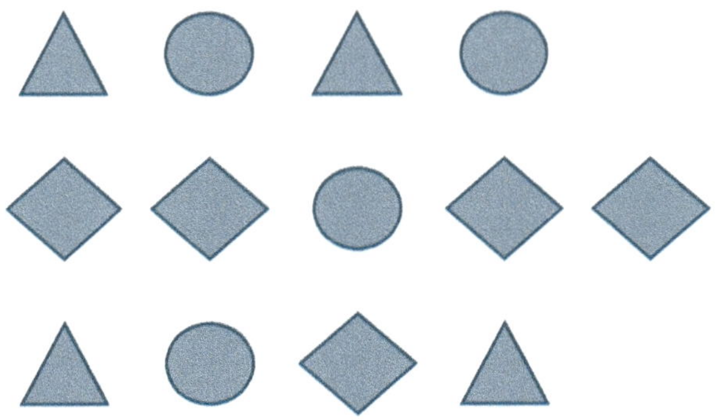

As children move on in their mathematical education, they begin to identify patterns in skip-counting, ten-frames, and hundreds charts. Many elementary students engage in calendar math, where they look for numerical patterns in days, weeks, months, and years on the calendar. As children progress into middle and high school, though, opportunities to identify patterns in studying mathematics seem to dwindle considerably.

Pause and Consider

In what ways could middle and high school students engage in identifying patterns when studying mathematics? Record your thoughts about this in your journal before reading further.

Middle and high school mathematics is fertile ground for students to continue to explore mathematical patterns. For example, you might give middle school students the task in Figure 3.6, which requires them to use their understandings of fraction operations to explore a numerical pattern.

Explore the sequence below to find a pattern in the fractions represented by each term.

$$\frac{1}{1+1} \ , \quad \frac{1}{1+\frac{1}{1+1}} \ , \quad \frac{1}{1+\frac{1}{1+\frac{1}{1+1}}} \ , \quad \cdots$$

Source: https://nzmaths.co.nz/

Similarly, you might give students studying linear relationships a task that asks them to explore patterns found by varying the slope of a line (see Figure 3.7).

Graph the following lines on the same coordinate plane using your graphing calculator:

1. $y = x$
2. $y = 2x$
3. $y = 3x$
4. $y = \frac{1}{2}x$
5. $y = \frac{1}{3}x$

What seems to be happening as the slope of $y = x$ gets larger? If the slope is between 0 and 1? Write a conjecture about the effect of the size of positive slopes on the graph of a line.

Even in calculus, students could engage in patterning tasks, such as the one in Figure 3.8, which asks them to identify patterns in logarithmic functions.

Pause and Consider

Now that we have provided a few examples of patterning tasks for middle and high school mathematics, think about the mathematical content that you teach. In your journal, list three or four ideas about how your students could explore mathematical relationships and identify patterns.

In the analytic framework for reasoning-and-proving, Stylianides (2010) describes two types of patterns: *plausible patterns* and *definite patterns*. In discussing plausible patterns, he presents a common pattern task that can be found in school textbooks (see Figure 3.9).

FIGURE 3.9 Common pattern task found in school textbooks.

What are the missing terms in the table below?

N	1	2	3	4	5	6
P	4	8	12			

Stylianides (2010) then explains,

> The tendency is to solve this task as if it concerned an arithmetic sequence. In that case, the pattern could be described by the algebraic expression $P = 4N$ and the missing terms would be 16, 20, and 24. This solution to the task is based on the *assumption* that the terms are part of an arithmetic sequence, but there is no basis for this assumption in the task. Indeed, there are an infinite number of patterns that will fit the given data set. For example, one could say that the first three terms are the "unit of repeat" of a pattern and so the missing terms are 4, 8, and 12. Someone else could provide a more complicated pattern that also satisfies the given set of data, for example, $P = \alpha N\ 3 - 6\alpha N^2 + (4 + 11\alpha) \times N - 6\alpha$, where α can be any real number. (p. 40)

This task, then, can produce **plausible patterns**, where "it is impossible to provide conclusive mathematical evidence for the selection of a specific pattern over alternative patterns that also fit the given data" (p. 40).

Consider the Toothpicks Task in Figure 3.10. Stylianides (2010) argues that this task gives rise to a **definite pattern** because the text of the task makes explicit that the figures continue to grow in a defined way, thus resulting in only one unique way to describe the mathematical relationship for the perimeter of the Nth figure. If that text were not present, then other patterns could arise. For example, if the text defining the growth of the shapes were not present, then Shapes 4, 5, and 6 could repeat Shapes 1, 2, and 3, establishing another defendable pattern. So, if that text were not there, then a number of *plausible patterns* could arise from this task.

Plausible Patterns: different possible patterns that can be considered as correct responses to the same patterning task because the text has not specified how the pattern continues.

Definite Pattern: the only correct response to a patterning task because the text has specified how the pattern continues.

FIGURE 3.10 The Toothpicks Task.

The Toothpicks Task

The shapes shown on the next page are made with sticks of equal sizes. Shape 1 consists of 4 toothpicks arranged to make a square. Shape 2 is made by adding a column of squares to the right end of Shape 1, with the column having one additional square than the previous column. The shapes continue to grow in this manner—for each subsequent shape, a column is added to the right that is one square more than the previous column.

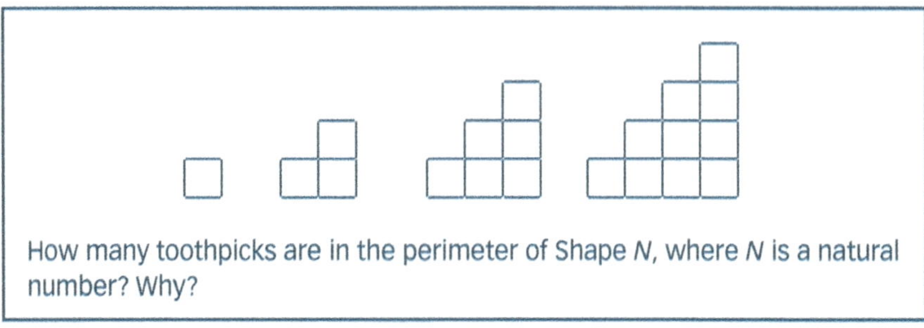

How many toothpicks are in the perimeter of Shape *N*, where *N* is a natural number? Why?

Source: Adapted from Stylianides (2010).

Making a Conjecture

Conjecture: a reasoned hypothesis about a general mathematical relation based on incomplete evidence.

Stylianides (2010) defines a ***conjecture*** to be "a reasoned hypothesis about a general mathematical relation based on incomplete evidence" (p. 41). He goes on to comment on the relationship between identifying a pattern and making a conjecture:

> The activities "making conjectures" and "identifying patterns" are clearly related as parts of the more general activity of "making generalizations." Yet, I identify an important distinction between the two: in conjecturing the solver formulates a hypothesis about a generalization whose domain extends beyond the domain of cases that the solver checked, whereas in pattern identification it is possible for the generalization to cover only the examined cases—Reid 2002. (p. 41)

Conjecturing requires that students go beyond describing a possible pattern to making a supposition about an entire set of numbers or objects.

Before we continue with the other parts of the framework, take a few minutes to engage in Activity 3.4, which asks you to complete the Toothpick Task.

Activity 3.4 Solve the Toothpick Task

1. Complete the Toothpick Task, contained in Figure 3.10.

2. Once you have identified a pattern between the shape number and the number of toothpicks in the perimeter of that shape, write a conjecture that generalizes the relationship.

3. Write an argument that justifies that your conjecture is true. Do you think your argument counts as a proof (refer to Figure 3.3)? Why or why not? Record your thoughts in your journal.

Debriefing Activity 3.4: Solve the Toothpick Task

One way to approach the Toothpick Task is by creating a table of values as shown in Figure 3.11.

FIGURE 3.11 A table of values for the Toothpick Task.

Shape Number (N)	1	2	3	4	5	...	N
Perimeter (P)	4	8	12	16	20	...	?

Reasoning from this table, you might see that as you move from one shape to the next, four toothpicks are added to the previous perimeter. This is looking at the situation as having a ***recursive relationship***—the perimeter of Shape N relies on knowing the perimeter of Shape $N - 1$. If you saw this relationship, you may have written something like this for your conjecture: "The perimeter of a shape is four more than the perimeter of the previous shape." If you made conjectures about the number of toothpicks in the perimeter of Figure N through inspecting a table of values, then you may have found it challenging to write a convincing argument that this is *the definite pattern* that arises from this situation—a convincing argument would need to make reference to the mathematical structure

Recursive Relationship: a mathematical definition in which the first case is given and the nth case is defined in terms of one or more previous cases and especially the immediately preceding one (www.merriam-webster.com).

of the task and the process by which each new shape is created. (We discuss proofs for the Toothpick Task in the next section on developing arguments.)

Alternatively, you might also have noticed that the value of the perimeter is four times the shape number. This is called the *explicit relationship*, where the perimeter is determined based on knowing the shape number. If you saw the explicit relationship, you may have written a conjecture that looked something like $P = 4N$ or written something like, "The perimeter of any figure is four times the figure number." We discuss additional conjectures for the Toothpick Task in the upcoming section on developing arguments.

Reflect back to the content of Chapter 2 and the limitations of using solely empirical examples in reasoning-and-proving activities. Arguing the "truth" of your conjecture about the perimeter of the toothpick shapes solely from the values in the table relies on the use of empirical examples. How do you know that the toothpick shape pattern doesn't fall apart between Shape 5 and Shape N (as the numerical pattern did in the Circle and Spots Problem)? How do you know that this pattern *always* holds, if the shapes continue to grow as described in the Toothpick Task? Students in middle and high school often use this kind of reasoning when presented with patterning tasks such as this one. We hope we have convinced you that this kind of reasoning based solely on empirical examples is limiting from a reasoning-and-proving perspective.

So, what are we to do to move our students beyond this kind of limited reasoning? We teach them to develop mathematical arguments and the differences between types of mathematical arguments. The Reasoning-and-Proving Framework (Figure 3.4) provides guidance in this area. In the next section, we describe the mathematical activities embedded in *developing arguments*.

Explicit Relationship: a functional relation in mathematics in which the dependent variable is stated directly in terms of the independent variable (www. merriam-webster. com).

DEVELOPING ARGUMENTS

Look back at Figure 3.4 and note that Stylianides (2010) describes two types of activities that occur when *developing arguments:* developing a proof and developing a non-proof argument. Each of these activities is then further delineated. There are two types of proof in this framework—demonstration proof and generic argument—and two types of non-proof arguments—empirical argument and rationale. We explain each type of argument below, beginning with the non-proof arguments.

Non-proof arguments. Stylianides (2010) defines a ***non-proof argument*** as "an argument for, or against, a mathematical claim that does not qualify as proof" (p. 42). He then describes an ***empirical argument*** as "an argument that purports to show the truth of a claim on the basis of the confirming evidence offered by examination of a proper subset of all the possible cases" (p. 42). Another way to think about empirical arguments is to consider how we defined them in Chapter 2, where we described an empirical argument as an argument that uses several examples of specific cases without ever moving from the specific to the general. For the Toothpick Task, an empirical argument might look like this: "I tried five shapes, and it looks like the perimeter of each shape is four times the shape number. So, it makes sense that the perimeter of Shape N is $4N$." In the set of student responses to the Sum of Two Odds task (Figure 3.1), Student E makes an empirical argument. In the vignettes from Hoover High School in Chapter 1, Shonda—the student in Carly Epson's vignette—makes an empirical argument.

Non-proof Argument: an argument for, or against, a mathematical claim that does not qualify as proof.

Empirical Argument: an argument that uses several examples of specific cases without ever moving from the specific to the general.

Pause and Consider

Think back to your practices as a teacher of mathematics. Have you accepted empirical arguments from your students as justification for mathematical claims? If you have, you are not alone. Knuth (2002b) observed that half of the secondary teachers he worked with accepted empirical arguments as proof and were convinced of the truth of a statement from considering only empirical examples. Record your thoughts in your journal about the extent to which you have accepted empirical arguments as justification for mathematical claims. Provide examples of the types of tasks your students were working on when they provided an argument based on empirical examples.

Rationales: valid arguments for, or against, mathematical claims that are not sufficiently developed to meet the standard of proof.

Stylianides (2010) defines ***rationales*** as "valid arguments for, or against, mathematical claims that are not sufficiently developed to meet the standard of proof" (p. 42). Student G's response (Figure 3.1) is a good example of a rationale. As we explained in Table 3.1, Student G makes a general argument about the addition of two odd numbers, but it relies on many things being known by the reader. In order to make this a proof, Student G needs to define odd and even numbers and provide more explanation regarding why the process of "circling by twos" makes sense in this situation.

DEVELOPING A PROOF

Stylianides (2010) writes,

> I define a *proof* to be a valid argument based on accepted truths for, or against, a mathematical claim. By "accepted truths" I mean axioms, theorems, definitions, and other statements that can comfortably be taken as shared and without justification in a particular classroom community at a given time. (p. 41, emphasis in original)

Demonstration Proof: a proof that begins from a generalized perspective and does not rely on the representativeness of a particular case.

When we think about a mathematical proof, it is likely that we envision a ***demonstration proof.*** This type of proof forms the backbone of high school geometry and is most prevalent in studying mathematics at the post-secondary level. In the set of student work in Figure 3.1, Student H provides the most commonly seen type of algebraic demonstration proof (shown again in Figure 3.12).

FIGURE 3.12 Student H's demonstration proof of the Sum of Two Odds task.

Student H

Definition of an even number: An integer p is even if and only if there is an integer k such that $p = 2k$

Definition of an odd number: An integer q is odd if and only if there is an integer k such that $q = 2k + 1$. Let's assume X and Y are odd where $X = 2n + 1$ and $Y = 2m + 1$, and n and m are integers.

Statement	Reason
X is an odd number so $X = 2n + 1$	Given, Definition of odd number
Y is an odd number, so $Y = 2m + 1$	Given, Definition of odd number
$X + Y = 2n + 1 + 2m + 1$	Addition Property of Equality
$X + Y = 2n + 2m + 1 + 1$	Commutative Property of Addition
$X + Y = 2n + 2m + 2$	Substitution
$X + Y = 2(n + m + 1)$	Distributive Property
$n + m + 1$ is an integer	Closure Property of Addition for Integers
$X + Y$ is an even number	Definition of an even number

Note that the demonstration proof starts from a generalized perspective—the use of variables to represent an entire class of numbers. Student H defines the variables to be used and builds a proof from those generalized definitions. Keep this in mind, as this characteristic will be an important distinction when understanding what differentiates a demonstration proof from a generic argument, which we define in the next section.

Pause and Consider

Before moving on, look back at the student responses in Figure 3.1, find other examples of demonstration proofs, and make note of those proofs in your journal.

Student F's response is another example of a demonstration proof because it begins from a generalized perspective. Note that Student F wrote the proof in paragraph form, which is different from Student H. What is similar between the two responses is that they both begin from a generalized perspective.

Although quite different in appearance, Student I's response is also considered a demonstration proof. This is a proof by exhaustion;

Student I described the entire class of odd numbers by focusing on the units digit—and thus covers all cases of the sum of two odd numbers, making the argument from a generalized perspective. Other types of demonstration proofs that you may know of include proof by mathematical induction, proof by contradiction, and proof by counterexample (used to disprove a conjecture). All of these types of proofs depend on beginning from a generalized perspective. Student J's proof is also an example of a demonstration proof.

Thus far, we have been able to describe three of the four proofs from the set of student responses in Figure 3.1 as demonstration proofs. This leaves Student A's response, which we argued is a proof because although the student begins with a specific example, the argument is expanded to cover all odd numbers. In addition, this response meets the four criteria for judging whether an argument is a proof (Figure 3.3). But this proof does not meet the description of a demonstration proof because it *begins with a specific example as opposed to beginning from a generalized perspective.*

Stylianides (2010) calls Student A's proof a **generic argument** and provides this definition: "a generic argument is a proof based on a particular case that is treated as a representative of the general case" (p. 41), as opposed to a demonstration proof, which "does not rely on the representativeness of a particular case" (p. 42).

FIGURE 3.13 Student A's response to the Sum of Two Odds task.

Student A

If I take the numbers 5 and 11 and organize the counters as shown, you can see the pattern. You can see that when you put the sets together (add the numbers), the two extra blocks will form a pair and the answer is always even. This is because any odd number will have an extra block and the two extra blocks for any set of two odd numbers will always form a pair.

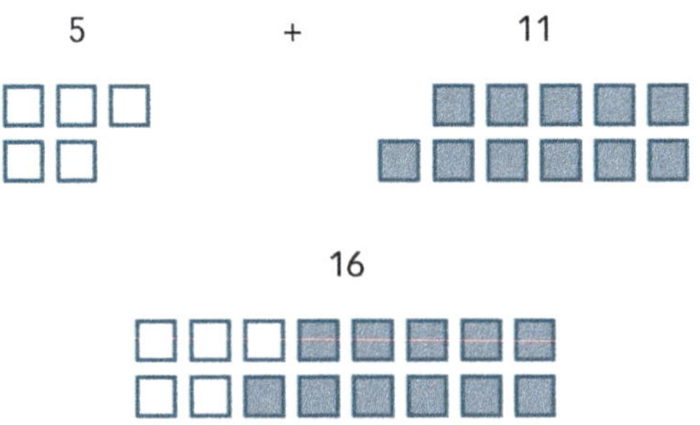

Source: Adapted from Coxford, et al. (2003).

Note that Student A begins with a numerical example (5 + 11 = 16) and, from that empirical example, argues that the mathematical structure of odd numbers—that all odd numbers will have an extra block—means that adding two odd numbers will always result in having two extra blocks, which can be paired together.

What would a generic argument look like for the Toothpick Task? Figure 3.14 contains two different generic arguments for this task. Although each argument produces different-looking algebraic expressions to represent the perimeter of the Nth figure (which are based on the thinking involved in developing the expression), note that the two expressions are equivalent.

FIGURE 3.14 Two generic arguments for the Toothpick Task.

In the fourth figure, you can move the tops of the columns (follow the arrows) to create a square with sides of 4. I know that the perimeter of a square is 4 times the side of the square, so the perimeter of this square is 4 × 4 or 16. 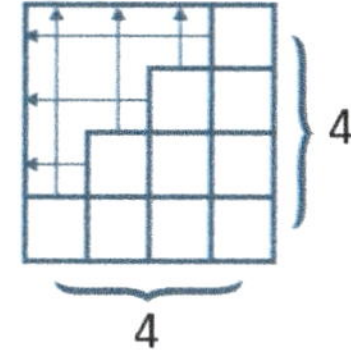 This will happen for any size figure—you can move the toothpicks to create an $n \times n$ square, which has a perimeter of $4n$. Thus, the formula $P = 4n$, where n is the figure number, works for any size figure.	In the fourth figure, the width is 4 and the height is 4. Then for each column, there are 2 toothpicks at the top that need to be counted, and there are 4 sets of those. So, the perimeter is $4 + 4 + 4(2) = 16$. 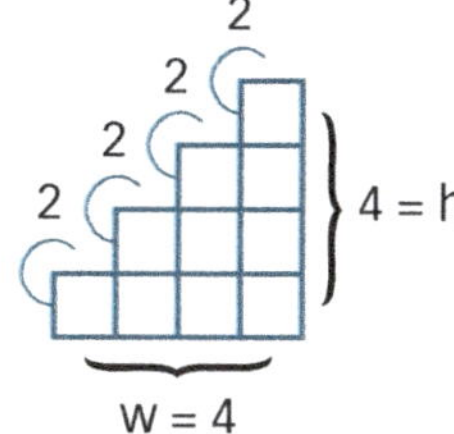 The same pattern is going to hold for any size figure. So, for the nth figure, the width = n and the height = n, and there will be n columns with 2 toothpicks at the top of each column. So, the equation for any size figure is $P = n + n + n(2)$.

Learning to build generic arguments has the potential to strengthen students' understanding of and capacity to engage in reasoning-and-proving. Building on Stylianides' (2008a, 2010) work, Karunkaran, Freeburn, Konuk, and Arbaugh (2014) wrote that generic arguments "provide powerful access for Grades 6–12 students to engage in reasoning-and-proving activities not typically found in middle and high school mathematics" (p. 159). Furthermore, they wrote,

> Consider, for example, algebra students finding patterns for the exponent rules using real numbers and then arguing that the patterns always hold in general, or geometry students finding patterns for the sum of the interior angles of polygons using specific polygons and then arguing that the pattern holds for any polygon. For many of the "rules" for mathematical relationships, Grades 6–12 students could begin by examining one or more empirical examples and then construct a mathematical argument based on those empirical examples showing that the pattern must exist for all cases, thus constructing a generic [argument]. Furthermore, we believe that generic [arguments] can provide scaffolding for students who are learning to write demonstration proofs in Grades 6–12. (pp. 159–160)

These authors also argue that when middle and high school students engage in developing generic arguments, they are engaging in three of the eight Standards for Mathematical Practice (SMP) as defined by the Common Core State Standards—Mathematics (NGA, CCSSO, 2010). The most obvious connection to the SMPs is *construct viable arguments and critique the reasoning of others*. Less obvious perhaps, but in the forefront of developing generic arguments, are the SMPs *look for and make use of structure* and *look for and express regularity in repeated reasoning*.

Teaching Takeaway:

When students write generic arguments, they engage in the SMPs *look for and make use of structure* and *look for and express regularity in repeated reasoning.*

Look back at Student A's response to the Sum of Two Odds task (see Figure 3.13). Student A used the *structure* of odd and even numbers in developing the generic argument. While Student A did not explicitly display *looking for regularity in repeated reasoning* in the response, we can imagine that Student A explored a number of empirical examples to identify a pattern in preparation for writing the generic argument, thus engaging in these two SMPs.

REFLECTING ON WHAT YOU'VE LEARNED ABOUT REASONING-AND-PROVING

In Chapter 2, you learned about the limitations of basing mathematical arguments on empirical examples. Earlier in this chapter, you created a list of criteria to use to judge whether a mathematical argument can count as a proof (see Figure 3.3) and then read about the Reasoning-and-Proving Framework, which provides information about reasoning-and-proving from a mathematical, learner, and pedagogical perspective (see Figure 3.4). In unpacking the components of the framework, you learned how to start with an empirical example to build a type of proof called *generic argument* and how generic arguments are different from *demonstration proofs*. You also learned about *plausible patterns* and *definite patterns, conjectures,* and *non-proof arguments* (*rationales* and *empirical arguments*). This is a lot of new information!

Learning about and understanding these ideas about reasoning-and-proving serves to form a strong foundation on which to build teaching practices that support students to better understand and be able to engage in reasoning-and-proving. Before we turn our attention to those teaching practices in the second half of this book, we encourage you to reflect on what you have learned about reasoning-and-proving thus far by responding to the prompts in this Pause and Consider and then engaging in Activity 3.5.

Pause and Consider

When learning a lot of new information, it is helpful to use writing to reflect on what you feel you've learned and what questions remain for you about the new information. To facilitate this reflection, respond to the following prompts in your journal. Refer back to pertinent parts of this chapter as needed.

- How confident are you that you could reliably use the criteria for judging whether an argument counts as a proof (Figure 3.3) to assess students' work to a reasoning-and-proving task? (You will have another opportunity to use the criteria in Chapter 6 to assess a set of student work.)

- Write a description of the difference between *plausible patterns* and *definite patterns*. How could you explain this to students?
- What is a conjecture? What kind of reasoning-and-proving activity supports students to make conjectures?
- What is an empirical argument? Why are empirical arguments limiting in the proof-writing process?
- What is a demonstration proof? A generic argument? How are they similar? What differentiates them?

In responding to these prompts, you may have noted that your understanding of the information we presented in this chapter is stronger in some areas than in others. If that is the case and you are working through this book with other teachers, then engaging in discussions about this content will help you to better understand ideas that you may not understand thus far. And although we turn to considering teaching practices that support reasoning-and-proving in your classroom for Chapters 4–6, we revisit many of these ideas along the way. Before we move on, though, we want to give you another opportunity to solidify your understandings of and capacities to generate generic arguments by revisiting the Squares Problem from Chapter 2.

REVISITING THE SQUARES PROBLEM FROM CHAPTER 2

When you worked on the Squares Problem in Chapter 2 (Activity 2.1), we did not ask you to write a proof of how many 3×3 squares there are in the generalized $n \times n$ square, although you may have conjectured at that time that the generalization for the number of 3×3 squares in an $n \times n$ square can be represented by $(n - 2)^2$, where n is the length of the side of the original square. Now that you had the opportunity to explore different types of proofs, the next activity asks that you write a generic argument for the Squares Problem.

Activity 3.5 Writing a Generic Argument for the Squares Problem

Return to your work for Activity 2.1 and write a generic argument for finding the number of 3 × 3 squares in an $n \times n$ square. Remember to start your argument with an empirical example and then argue that this example is representative of the mathematical relationships (structure) that exist in all cases.

Debriefing Activity 3.3: Writing a Generic Argument for the Squares Problem

Teachers with whom we have worked have generated a number of different generic arguments for the Squares Problem. Each argument is based on a pattern that the teachers identified when working with empirical examples. Here are two of the generic arguments that have been most prevalent.

Generic Argument #1. This generic argument builds on the explanation made in Chapter 2 for calculating the number of 3 × 3 squares in a 60 × 60 square, which we repeat in this generic argument.

So rather than beginning in the lower left corner and actually moving the 3 × 3 square along the lower portion of the 60 × 60 square, we need to consider how many 3 × 3 squares would fit horizontally across the lower portion of the larger square (see Figure 3.15). If we were moving a 3 × 1 rectangle instead of a 3 × 3 square, we could fit sixty 3 × 1 rectangles horizontally across the bottom row of the 60 × 60. But since we are moving three columns, not one, we will be able to fit only 58 (or 60 – 2) 3 × 3 squares. If we move the 3 × 3 square up vertically, we will note that for each row we move up, there are 58 possible 3 × 3 squares that will fit horizontally. When we finish this process, we will have 58 × 58, or $(60 - 2)^2$, different 3 × 3 squares.

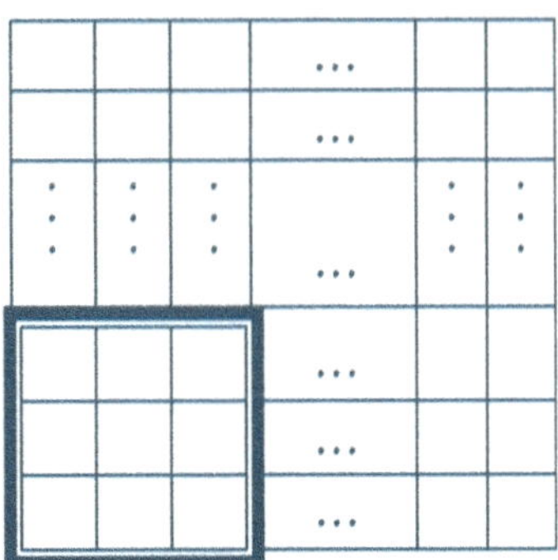

FIGURE 3.15 Picture for the 60 × 60 square.

Building on this argument for a 60 × 60 square, we can see that the same pattern will hold for an $n \times n$ square. The number of 3 × 3 squares that will fit both horizontally and vertically is $(n - 2)$, so the total number of squares in an $n \times n$ square is $(n - 2)(n - 2)$ or $(n - 2)^2$.

Generic Argument #2. Figure 3.16 contains a different generic argument for how many 3 × 3 squares there are in an $n \times n$ square.

FIGURE 3.16 Generic Argument #2 for the Squares Problem.

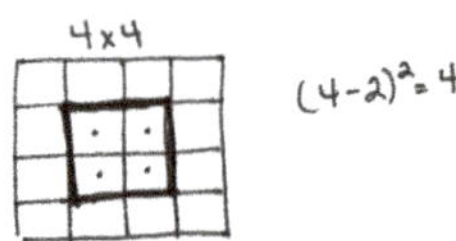

Each 3x3 square must have a unique center "block" In order for the 3x3 square to be contained within the nxn square, the center block cannot occur on the edge of the nxn square. Since the center block cannot occur on the edge of the nxn square, we are always left with the remaining (n-2) by (n-2) square. Thus since each center block represents a unique 3x3 square, there are (n-2)² 3 by 3 squares in an nxn square.

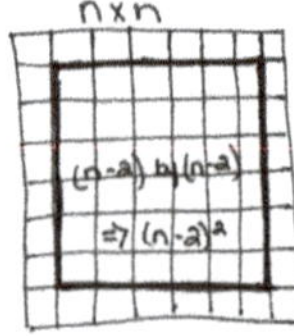

WE REASON & WE PROVE FOR ALL MATHEMATICS

Being able to construct both of these generic arguments depends on using empirical examples (i.e., drawing and counting the number of 3 × 3 squares in 4 × 4, 5 × 5, and 6 × 6 squares); looking for a pattern that arose from developing a strategic counting method; and then arguing that the counting method would work for any size square based on the structure of the $n \times n$ square and the 3 × 3 squares. Often in middle and high school, we expect students to jump directly to writing a proof or mathematical generalization for a mathematical relationship when they have not spent time exploring through empirical examples. The exploration phase (identifying patterns and writing conjectures) is a critical stage for students. In addition, and in the context of the Squares Problem, middle and high school students would have great difficulty writing a proof for the number of 3 × 3 squares in an $n \times n$ square without having access to generic arguments. (This can also be proven through mathematical induction, a proof method that is outside of the scope of middle and high school mathematics.)

Pause and Consider

What questions remain for you about the content of this chapter? Make note of these questions in your journal. Return to these questions as you work through the remaining chapters of this book in order to assess new understandings.

In this activity, you will build on what you have learned in this chapter to select a mathematical task that supports students to engage in the types of activities described in the Reasoning-and-Proving Framework (Figure 3.4).

CONNECTING to Your CLASSROOM

1. Select one of the classes you teach as the focal class for enacting this classroom connection. (If you are not currently teaching, consider a class you have taught in the past or will teach in the future.)

2. Identify a task that has the potential to engage students in some activity related to reasoning-and-proving. The task could come from your textbook or some other source, or you could create it yourself.

continued>>

<<continued

3. Solve the task as many ways as you can.

4. In your journal, respond to these questions:

   a. In what reasoning-and-proving activities will students engage through their work on the task?

   b. What prior knowledge and experiences should students have in order to successfully complete this task?

   c. What different approaches could students take in completing this task?

If you are working through this book with other teachers, be prepared to share your task and the responses to Question 4 above at your next meeting. We will also revisit this work at the end of Chapter 4.

*****

Chapters 1–3 have focused on supporting you to understand more deeply what reasoning-and-proving is and what kinds of thinking are involved when reasoning-and-proving. You have had a few glimpses into classrooms where the teachers implemented reasoning-and-proving activities with their students, and you began a page in your journal titled, "Pedagogical Moves That Support Students to Reason-and-Prove." We encourage you to go back to that list now and add other moves that have come to light in subsequent chapters. We now explicitly turn our attention to teaching practices that support reasoning-and-proving. In Chapter 4, you will analyze two narrative cases through the lens of four of the mathematics teaching practices found in *Principles to Actions: Ensuring Mathematical Success for All* (NCTM, 2014).

Discussion Questions

1. Using the criteria for judging when an argument counts as a proof (Figure 3.3), critique the argument you created in response to Discussion Question 1 in Chapter 2 (the Sum of Three Consecutive Numbers task). Does your argument meet the criteria for proof? Why or why not?

2. Using the Reasoning-and-Proving Framework presented in Figure 3.4, how would you classify the argument you created for the Sum of Three Consecutive Numbers task?

3. Consider these two tasks:

   - Prove that when you add any two odd numbers together, your answer is always even.
   - Prove whether the following statement is true or false: The sum of three consecutive numbers is divisible by 3.

Discuss alternative ways you could present each task that would provide students with the opportunity to engage in the wider range of reasoning-and-proving activities identified in Figure 3.4.

Helping Students Develop the Capacity to Reason-and-Prove

Ambitious teaching engages students in challenging tasks and provides appropriate levels of support to ensure that each and every student succeeds in meaningful mathematics learning. Such teaching—which encourages and promotes multiple solution paths—can be intimidating and exciting at the same time. Exciting because such tasks provide insights into what students are thinking, which can then be used to guide students forward. Intimidating because such tasks can be unbounded, resulting in a wide range of responses that the teacher may not be prepared to consider. Making reasoning-and-proving a central feature of mathematics instruction in secondary classrooms will require ambitious teaching and the use of challenging tasks. In this chapter, you will investigate a set of effective teaching practices that support such instruction. In particular, in this chapter you will be asked to consider

- how the effective teaching practices can be used to engage students in reasoning-and-proving in secondary mathematics classrooms; and
- how advancing planning, with attention to the effective teaching practices, can improve the quality of instruction.

HOW DO YOU HELP STUDENTS REASON-AND-PROVE?

In Chapters 1–3, you engaged in activities designed to enhance your understanding of what it means to reason-and-prove. This understanding is valuable to you as a teacher in helping your students develop the capacity to reason-and-prove. For example, knowing when an argument counts as a proof would have helped Barbara Law (from the vignette in Chapter 1) in recognizing the validity of Marissa and Michael's argument. There are, however, other teaching dilemmas, such as those faced by Carly Epson and Jason Steiner (from the vignettes in Chapter 1), that require the use of teaching strategies that help students move beyond the challenges they face in creating mathematical arguments.

This chapter is designed to begin to address the question of how to support students' capacities to reason-and-prove without taking over the thinking for them. We begin the chapter with a framework for examining mathematics instruction through a reasoning-and-proving lens, drawing on the effective teaching practices articulated in *Principles to Actions: Ensuring Mathematical Success for All* (NCTM, 2014). In particular, in this chapter we focus on four of the eight identified practices:

- Implement tasks that promote reasoning and problem solving,
- Use and connect mathematical representations,
- Pose purposeful questions, and
- Facilitate meaningful discourse.

These teaching practices provide a lens for considering the actions that a teacher takes prior to and during instruction that support students' capacity to reason-and-prove.

Chapter 4 builds on the content of Chapters 2 and 3 by affording glimpses, through two narrative cases, into classrooms where students are engaged in reasoning-and-proving activities. In addition, the criteria for judging whether an argument is a proof discussed in Chapter 3 can be used to judge the quality of the arguments produced by students who appear in the narrative cases featured in this chapter. Finally, by carefully analyzing the narrative cases, you have the opportunity to consider what students seem to be learning about mathematical content and reasoning-and-proving processes along with how this learning may benefit students both immediately and long term.

What challenges do you foresee in attempting to integrate reasoning-and-proving into your own classroom practice? Record your thoughts in your journal. If possible, discuss these challenges with colleagues. We will return to these challenges at the end of the chapter.

A FRAMEWORK FOR EXAMINING MATHEMATICS CLASSROOMS

Principles to Actions: Ensuring Mathematical Success for All (NCTM, 2014) identifies a set of eight research-based teaching practices that provide a framework for improving the quality of mathematics teaching and learning. These eight effective teaching practices (shown in Figure 4.1) describe the core activities in which teachers must engage in order to ensure the learning of each and every student. (For additional explication of the effective teaching practices, see Boston, Dillon, Smith, & Miller, 2017; Huinker & Bill, 2017; NCTM, 2014; Smith, Steele, & Raith, 2017.)

While it is essential for a teacher to thoughtfully consider each of these practices in preparing for and teaching a lesson, in this chapter we

FIGURE 4.1 Eight effective mathematics teaching practices.

- Establish mathematics goals to focus learning.
- **Implement tasks that promote reasoning and problem solving.**
- **Use and connect mathematical representations.**
- **Facilitate meaningful mathematical discourse.**
- **Pose purposeful questions.**
- Build procedural fluency from conceptual understanding.
- Support productive struggle in learning mathematics.
- Elicit and use evidence of student thinking.

Source: NCTM (2014)

focus our attention on the subset of practices highlighted in bold print in Figure 4.1. We have selected these four practices because of their centrality to engaging students in reasoning-and-proving. In the sections that follow, we describe each practice and what it means in the context of reasoning-and-proving. We provide an explicit discussion of "establish mathematics goals to focus learning" in Chapter 5. Implicit in our discussions in this chapter is the need to be clear about what you want students to learn as a result of engaging in a particular lesson.

Implement Tasks That Promote Reasoning and Problem Solving

According to *Principles to Actions: Ensuring Mathematical Success for All* (NCTM, 2014), "Effective teaching of mathematics engages students in solving and discussing tasks that promote mathematical reasoning and problem solving and allow multiple entry points and varied solutions" (p. 17). Such tasks require students to determine a course of action and chart a pathway to a solution because no solution strategy is implied or suggested by the task. These tasks have been referred to as high-level or cognitively demanding because they require students to think, reason, and make sense of a situation (Smith & Stein, 1998; Stein, Grover, & Henningsen, 1996). These tasks have also been referred to as "low threshold high ceiling" tasks (McClure, Woodham, & Borthwick, 2011)—that is, tasks that all students can enter and explore at some level and, at the same time, that have the potential to engage students in challenging mathematics. These types of tasks have also been called "low floor high ceiling" tasks.

Take, for example, the Building a Staircase task shown in Figure 4.2, which can be characterized as a high-level task (see NCTM, 2014, p. 18 for characteristics of high-level tasks). There is no previously learned rule or procedure that can be used to solve this task. Students must determine the underlying structure of the pattern and use this structure to determine the number of steps in any staircase. The first two questions in the task provide support for entry into the task, helping students get a foothold on the problem. Students might begin the task by drawing staircases on grid paper, building staircases using cubes, or making a table that shows the number of blocks required for making staircases with different numbers of steps. These entry

Charles is planning to build a staircase from his yard to the deck on the back of his house. He knows that a one-step staircase takes one block to build, a two-step staircase takes three blocks, and a three-step staircase takes six blocks. He is not sure yet how many steps he will need, so he wants to figure out how many blocks might be needed for staircases with different numbers of steps.

1. How many blocks will Charles need for a five-step staircase? Six-step staircase? Seven-step staircase?
2. Determine how many blocks Charles will need for a 10-step staircase without building or drawing the staircase.
3. Write a rule for determining the number of blocks that Charles would need for a staircase with any number of steps.
4. Create an argument that could convince Charles that your rule will always work. You can use numbers, symbols, models, and/or pictures.

points can help students begin to see patterns and make conjectures about the relationship between successive staircases and between the number of steps in the staircase and the number of blocks required to build the staircase.

As a result of engaging in a lesson built around this task, a teacher might want students to understand the following:

1. An equation can be written that describes the relationship between two quantities.

2. For an argument to be a proof, it must show that it is true for all cases.

3. Proofs can utilize different representations, and the representations can be connected.

4. Using only numerical examples as the basis of an argument does not constitute a proof.

These statements form the basis for the learning goals for the lesson.

Tasks that promote reasoning and problem solving, such as the Building a Staircase task, can also be solved in different ways. Students can solve the Building a Staircase task by drawing a picture or building a model of a staircase (as shown in Figure 4.3); using the

Solutions	Symbolic Generalization	Type of Proof	Representations
Student A View the blocks as half of a rectangle. Then make two sets of stairs that combine to form the rectangle. One set of stairs has a width (or height) of n blocks. When connected to the other set of stairs, we have a width (or height) of n+1. So the area of the rectangle is n(n+1). Since we only want the number of the blocks for half of the rectangle, we divide the area by 2. Hence, the number of blocks needed to build a staircase of height n is $\frac{n(n+1)}{2}$	$\dfrac{n(n+1)}{2}$	Demonstration	• Picture • Written Language
Student B For a staircase of height n (n=3) Square the sides ⟹ n^2 Square (3^2) Add another column of n squares n^2+n rectangle (3^2+3) Divide the (n^2+n) rectangle by 2 in order to obtain the number of blocks in a staircase of height n $\dfrac{n^2+n}{2}$		Generic Argument	• Picture • Written Language
Student C I know that the area of a triangle is $\frac{1}{2}bh$. So, since the base and height of this △ are both n, the area is $\frac{1}{2}n \cdot n$ or $\frac{1}{2}n^2$. But I still have $\frac{1}{2}$ of the edge pieces to add in, so the whole formula is $\frac{1}{2}n^2 + \frac{1}{2}n$	$\dfrac{1}{2}n^2 + \dfrac{1}{2}n$	Demonstration	• Picture • Written Language

Solutions	Symbolic Generalization	Type of Proof	Representations
$S = 1 + 2 + 3 + \cdots + n$ ← List numbers to be added. $S = n + (n-1) + (n-2) + \cdots + 1$ ← List numbers in reverse. $2S = [1+n] + [2+(n-1)] + [3+(n-2)] + \cdots + [n+1]$ ← Add the two lists together. $2S = (1+n) + (1+n) + (1+n) + \cdots + (1+n)$ ← Simplify $2S = n \cdot (1+n)$ ← Combine like terms: n groups of (1+n). $S = \dfrac{n \cdot (1+n)}{2}$ ← Divide both sides by 2.			

picture or model to identify a familiar shape (a rectangle, a square, and a triangle, respectively); and then connecting the number of blocks in the shape to the number of steps in the staircase. Or some students might employ a more numeric/algebraic approach, as shown in Student D's solution in Figure 4.3.

Students might also solve the Building a Staircase task by constructing a table of values as shown in Figure 4.4. A student might recognize that the relationship is not linear since the first difference—the difference in any two successive values of b when n is incremented by 1—is not constant. This could lead to finding the second difference and concluding that the relationship must be quadratic since the second difference is constant. From here, the student can use the general form of a quadratic equation $f(x) = ax^2 + bx + c$ and a series of substitutions, using points from the table, to determine the equation. (For a complete description of how to find a quadratic equation from a table of values, see http://sciencing.com/quadratic-equations-table-10001169.html.) While this strategy leads to a correct equation that shows the relationship between the two variables (which was Goal 1 for this task—an equation can be written that describes the relationship between two quantities), it does not help a student create a convincing argument and makes no connection to the context of the problem. Too often, students rush to create tables and find equations from naked numbers without using the context to make sense of the situation, thus limiting their engaging in reasoning-and-proving. In a problem such

FIGURE 4.4 A table representing the number of steps and corresponding number of blocks in the Building a Staircase task.

Number of Steps (n)	Number of Blocks (b)
1	1
2	3
3	6
4	10
5	15

as the Building the Staircase task, the context should be central to creating an argument. (The use of context is discussed in more detail in Chapter 6.)

While good reasoning-and-proving tasks must engage students in reasoning and problem solving, not all tasks that engage students in reasoning and problem solving foster reasoning-and-proving. Tasks that engage students in reasoning-and-proving must also provide opportunities for students to make generalizations (i.e., identify patterns and make conjectures) and develop arguments (i.e., proof and non-proof) (Stylianides, 2010) as we described in Chapter 3.

Teaching Takeaway:

Good reasoning-and-proving tasks must require students to reason, problem solve, make generalizations, and develop arguments.

The Building a Staircase task provides students with opportunities to engage in all of these processes. For example, after building or drawing several staircases, students might notice specific patterns, such as the number of blocks in a staircase is the sum of the numbers 1 to n, where n is the number of steps. For example, a staircase with three steps has columns of $1 + 2 + 3$ blocks; a staircase with four steps has columns of $1 + 2 + 3 + 4$ blocks. As previously mentioned, students might see that each staircase can be viewed as a triangle plus edge pieces (Student C), a large square plus another column of squares (Student B), or two identical staircases fit together to form a rectangle (Student A). Question 3 asks students to write a rule to describe the number of blocks in any staircase, which could result in several different but equivalent generalizations as represented by the algebraic expressions in the second column of Figure 4.3. This could also be written in a more descriptive form without symbolic notation. Finally, Question 4 explicitly asks students to create a convincing argument that could take the form of a generic argument (Student B) or a demonstration proof (Students A, C, and D). (Refer back to Chapter 3 for a full discussion of these two types of proofs.)

Although the Building a Staircase task provides the opportunity for students to engage in the full range of reasoning-and-proving processes, it is not necessary for every task to do so. Students need opportunities to build their capacity for creating sound arguments by first having opportunities to identify patterns and make conjectures. According to Stylianides (2008a):

> It is thus important that students be assisted to develop proficiency in, and understand the relations among, all major activities that are frequently part of the process of making sense of and establishing mathematical knowledge: "identifying patterns," "making conjectures," "providing non-proof arguments," and "providing proofs." (p. 9)

Use and Connect Mathematical Representations

According to *Principles to Actions: Ensuring Mathematical Success for All* (NCTM, 2014), "Effective teaching of mathematics engages students in making connections among mathematical representations to deepen understanding of mathematics concepts and procedures as tools for problem solving" (p. 24). As shown in Figure 4.5, students can represent mathematical ideas using visuals, symbols, language, contexts, and models. Take a minute and look back at the student work for the Sum of Two Odds problem (Figure 3.1). While Student E used a picture to communicate the solution, you can imagine that this student had access to physical squares (manipulatives) while exploring patterns. Student H and Student I both used symbolic representations, with Student H using variables (algebraic representation) and Student I using numbers (numeric representation). Student F used written language in the solution. Although not represented in the student work in Figure 3.1, in Figure 4.5 we provide an example of what a visual representation might be for the Sum of Two Odds task, with the "sticks" representing units and two circled sticks representing the number 2. It is important to note that not all tasks can be solved with any of the five representations. The Sum of Two Odds task does not lend itself to a contextual representation (a real-life or fictional situation that

embodies the mathematics). In Chapter 6, we focus more on the role of context in reasoning-and-proving.

Bringing your attention back to Figure 4.5, the double-headed arrows between each pair of representations shows that you can start with either representation and relate it to the other. Research suggests that good problem solvers are able to move flexibly between and among different representational forms and that strengthening these translation abilities facilitates learning (Lesh, Post, & Behr, 1987).

FIGURE 4.5 Different ways to represent a mathematical idea and the connections between them.

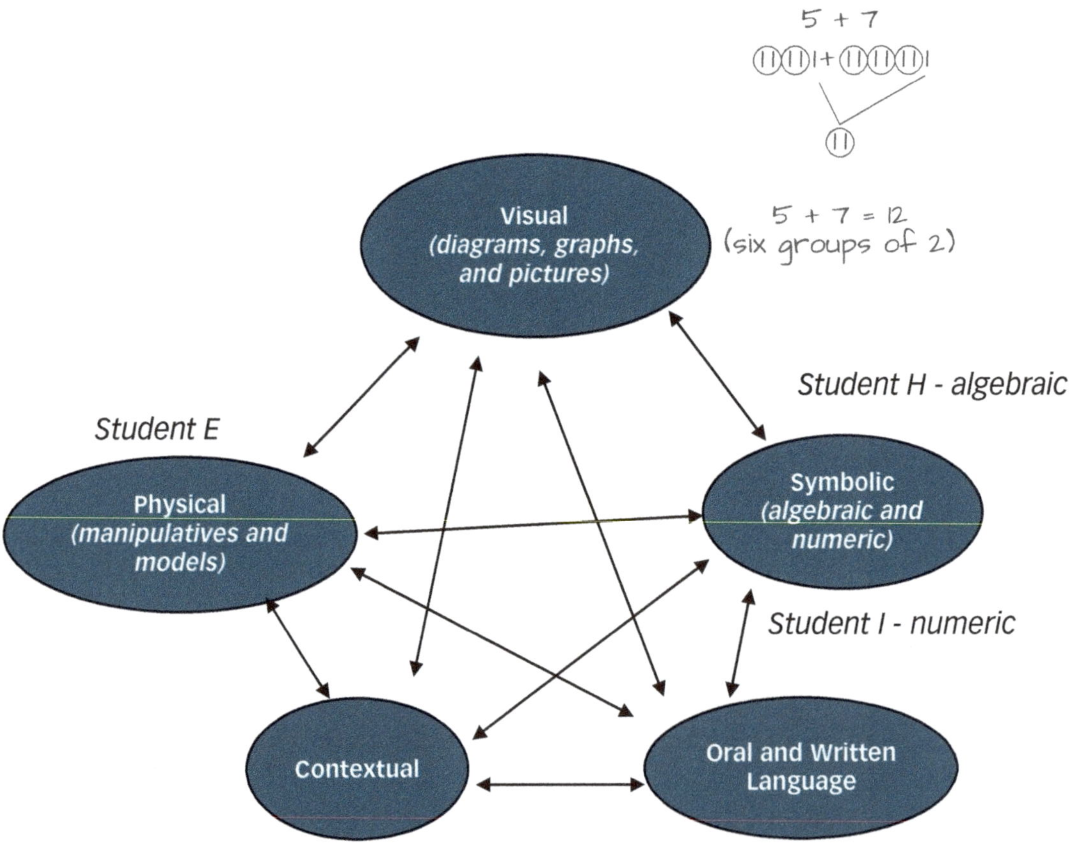

Source: Adapted from NCTM, 2014.

   WE REASON & WE PROVE FOR ALL MATHEMATICS

In the context of reasoning-and-proving, representations are particularly important. As we discussed in Chapter 3, proofs are often conceptualized in a two-column format labeled "statements" and "reasons" (see Student H in Figure 4.6, for example). Because of this, proofs have often been reduced to a set of symbols and verbal descriptions (two of the representations) and, while valid, they provide limited insight into why a mathematical statement must be true—they have limited *explanatory power*. According to Nelson (1997), "Pictures or diagrams can help the reader see why a particular mathematical statement is true and also to see how to begin to go about proving it true" (back cover). So, particularly in middle and high school mathematics, students should strive to write proofs with high explanatory power, and visual and physical representations are useful tools for writing a proof with high explanatory power. In proving that the sum of two odd numbers is always even in Chapter 3, the model in the generic argument in Student E's work (see Figure 4.6) communicates why this conjecture is true in a much more compelling way than the two-column proof used in student work sample H.

Consider the representations that students used in solving the Building a Staircase task (Figure 4.2). The drawings (visual representations) used by Students A, B, and C provide insight into *why* the number of blocks in a staircase with n steps can be described by the corresponding algebraic expression (symbolic representation). For example, Student A could clearly "see" that putting together two sets of stairs with the same number of steps formed a rectangle with

Explanatory Power: the extent to which an argument makes visible the mathematics and the reasons for why the argument proves (or disproves) the conjecture.

Teaching Takeaway:

Using visuals and models in reasoning-and-proving helps students to write proofs and to read and understand others' proofs. Using different representations in proofs supports students to construct viable arguments and critique the reasoning of others (SMP3 from CCSSM).

FIGURE 4.6 Two solutions to the Sum of Two Odds task (Figure 3.1, p. 41 and 40).

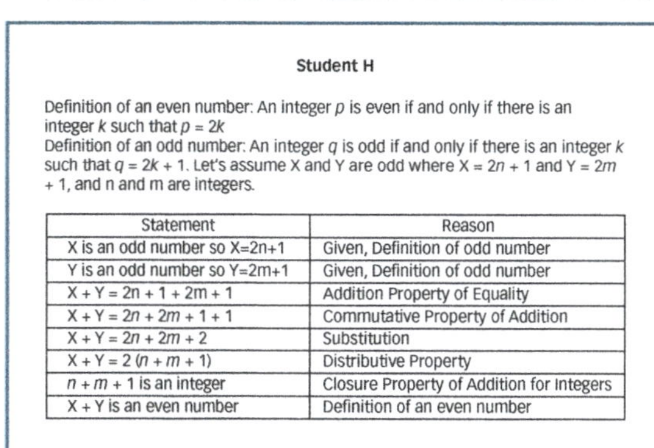

Student H

Definition of an even number: An integer p is even if and only if there is an integer k such that $p = 2k$
Definition of an odd number: An integer q is odd if and only if there is an integer k such that $q = 2k + 1$. Let's assume X and Y are odd where $X = 2n + 1$ and $Y = 2m + 1$, and n and m are integers.

Statement	Reason
X is an odd number so X=2n+1	Given, Definition of odd number
Y is an odd number so Y=2m+1	Given, Definition of odd number
$X + Y = 2n + 1 + 2m + 1$	Addition Property of Equality
$X + Y = 2n + 2m + 1 + 1$	Commutative Property of Addition
$X + Y = 2n + 2m + 2$	Substitution
$X + Y = 2 (n + m + 1)$	Distributive Property
$n + m + 1$ is an integer	Closure Property of Addition for Integers
X + Y is an even number	Definition of an even number

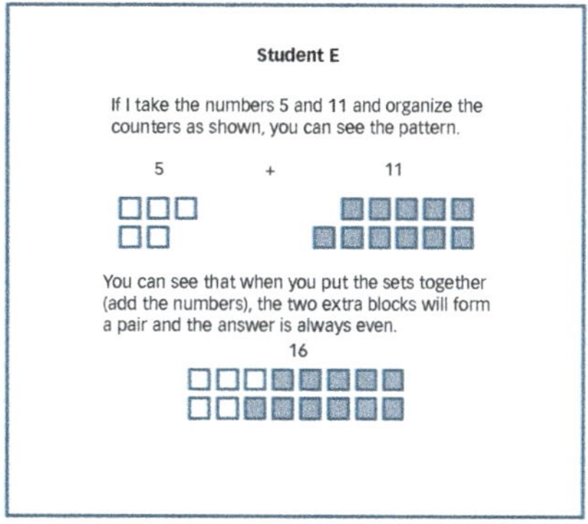

Student E

If I take the numbers 5 and 11 and organize the counters as shown, you can see the pattern.

You can see that when you put the sets together (add the numbers), the two extra blocks will form a pair and the answer is always even.

Source: Adapted from Coxford, A. F. et al (2003).

dimensions $n \times (n + 1)$, and so the number of blocks in one staircase would be half the area of the rectangle. Hence, "students are convinced of the conclusion's truth, not solely because of the deductive mechanism … but by the nature of the geometric pattern" (Knuth, 2002a, p. 488).

Students who recognized that the number of blocks in a staircase with n steps is the sum of the numbers 1 to n may look for a way to add the numbers together. This could lead to a method such as the one used by Student D in Figure 4.2. While this strategy results in a proof, it has no connection to the physical arrangement of the blocks needed to build a staircase. Having a whole-class discussion that includes connecting the mathematics in Student A's and Student D's solutions (both of which resulted in the same symbolic generalization) would help the students who produced these two solutions, as well as the rest of the students in the class, see that adding the same set of numbers twice (Student D) was the same as replicating the staircase (Student A)— one solution method was visual and the other was numeric. Making explicit connections among solution strategies that use different representations helps to develop students' understanding of mathematics and their flexibility among multiple representations.

Hence, the Building a Staircase task begins with a contextual representation (Charles is building a staircase to his deck) that can then be related to a visual representation or a physical model of the staircase (students could use multilink cubes to explore patterns with different-size staircases) and to a numeric representation. They all conclude with a symbolic representation (the algebraic expression) because that is what the task asked students to generate. The key to supporting student learning through the use of this task is not simply producing different representations and acknowledging that the problem can be solved in different ways, but ensuring that students understand the ways in which the models, drawings, and numeric/algebraic approaches all represent the problem context and can be related to each other.

Pose Purposeful Questions

According to *Principles to Actions: Ensuring Mathematical Success for All* (NCTM, 2014), "Effective teaching of mathematics uses purposeful questions to *assess* and *advance* students' reasoning and sense making about important mathematical ideas" (p. 35, emphasis added).

Asking assessing questions helps to elicit students' current understanding about a task (Smith, Bill, & Hughes, 2008). Specifically, assessing questions can help you determine what students are currently thinking and what they understand, gauge students' progress on the task, and inform next steps you can take to support the students' progress toward the lesson goals. In addition, assessing questions can support students' reflections about their approaches to, and mathematical thinking about, the task they are working on (Freeburn, 2015). Asking advancing questions helps students move from their current thinking toward the mathematical learning goals of the lesson (Smith et al., 2008). Specifically, advancing questions can support students to think about new or different representations; engage in mathematical processes (e.g., representing, generalizing, justifying); and think about new mathematical relationships and/or meanings (Freeburn, 2015). Figure 4.7 contains the characteristics associated with assessing and advancing questions.

While assessing questions are generally based closely on the work that a student has produced, the initial question(s) that a teacher asks may be

FIGURE 4.7 Characteristics of assessing and advancing questions.

Assessing Questions	Advancing Questions
• Are based closely on the work the student has produced	• Use what students have produced as a basis for making progress toward the target goal of the lesson
• Clarify what the student has done and what the student understands about what he or she has done	• Move students beyond their current thinking by pressing students to extend what they know to a new situation
• Give the teacher information about what the student understands	• Press students to think about something they are not currently thinking about
• Support students to be metacognitive	• May prompt students to consider new representations or mathematical relationships
Teacher stays to hear the answer to the question.	*Teacher walks away, leaving students to figure out how to proceed.*

Source: Adapted from Smith et al. (2017).

what Freeburn and Arbaugh (2017) have referred to as "hip pocket" questions. These are more general questions that the teacher can ask to begin a conversation with any student, such as, "Can you tell me what you did here? Can you explain how you are thinking about the problem so far?" Such questions provide a platform for the student to explain what he or she has done and send the message to the student that the teacher is interested in his or her thinking. Other assessing questions should then press the student to clarify parts of his or her explanation, thus letting the teacher know exactly what the student understands and how the student is thinking about the mathematics.

Consider, for example, Student D's solution to the Building a Staircase task, which we present again in Figure 4.8. The assessing questions in column 2 show a progression from a general "hip pocket" question to questions that are closely connected to the work the student has actually produced. Answers to these questions will make it clear to the teacher *exactly* what the student understands about the situation. As a teacher, it is important for you to ask assessing questions, even

FIGURE 4.8 Possible assessing and advancing questions for Solution 4 to the Building a Staircase task.

Student D	Assessing Questions	Advancing Questions
$S = 1 + 2 + 3 + \cdots + n$ ← List numbers to be added. $S = n + (n-1) + (n-2) + \cdots + 1$ ← List numbers in reverse. $2S = [1+n] + [2+(n-1)] + [3+(n-2)] + \cdots + [n+1]$ ← Add the two lists together. $2S = (1+n) + (1+n) + (1+n) + \cdots + (1+n)$ ← Simplify $2S = n \cdot (1+n)$ ← Combine like terms: n groups of $(1+n)$. $S = \dfrac{n \cdot (1+n)}{2}$ ← Divide both sides by 2.	• Can you explain what you did here and why you did it? • What does n represent? What do $n - 1$, $n - 2$, $n - 3$, etc. represent? • Why did you reverse the numbers and add the two sets together?	• Will this formula work to find the sum of any set of numbers? How do you know? • Another student (Student A) came up with the same generalization and used a diagram to explain it. How does your solution relate to the more visual solution?

if you think you understand what the student did, since your understanding of the work may not be the same as the student's, and it is only through a student's explanation that the true meaning of the work is revealed.

Advancing questions can be asked only after you have assessed what the student knows. You can then consider what the student currently understands in light of what you ultimately want him or her to understand (the lesson learning goals) and ask questions that will move the student on a trajectory toward the learning goals of the lesson. Returning to Student D's solution in Figure 4.8, let's assume that the student was able to appropriately answer the teacher's assessing questions. The student's explanation may have started with the recognition that the number of blocks needed for a staircase with n steps was the sum of the numbers from 1 to n, so he or she needed to find an efficient way to add the numbers. Armed with this information, the teacher might ask one (or both) of the advancing questions shown in column 3 of Figure 4.8. These questions press the student to explain whether the generalization works for all cases (which was the teacher's learning goal #2—for an argument to be a proof, it must show that it is true for all cases) and to connect the representation of the generalization to a more visual representation (which was the teacher's goal #3—proofs can utilize different representations and the representations can be connected). A critical pedagogical move at this point is for the teacher to walk away and give the student (or group) time for further exploration.

Pause and Consider

Why might it be a "critical" pedagogical move for you to walk away after asking advancing question(s)? Record your thoughts in your journal. We will return to these thoughts at the end of this chapter.

Although you may ask assessing and advancing questions during instruction, it is helpful to develop questions prior to the lesson. By first anticipating how students might solve a task (correctly and incorrectly), you can then consider the questions that you can ask the student who produced a specific solution to assess and advance the student's understanding. This is critical, since, according to Smith and Stein (2018),

> Developing questions only "in the moment" is very challenging for a teacher who is juggling the needs of a classroom full of learners who need different types and levels of assistance. When teachers feel overwhelmed by the needs and frustrations of their students, it is easy for them to revert to just telling students what to do when an alternative course of action does not immediately come to mind. (p. 54)

Assessing and advancing questions are essential tools for effective mathematics teaching regardless of the content and processes that are the focus of a particular lesson. When engaging students in reasoning-and-proving tasks, assessing and advancing questions provide you with a way to make student thinking public so that you can guide and challenge it.

Facilitate Meaningful Mathematical Discourse

According to *Principles to Actions: Ensuring Mathematical Success for All* (NCTM, 2014), "Effective teaching of mathematics facilitates discourse among students to build shared understanding of mathematical ideas by analyzing and comparing students' approaches and arguments" (p. 29). Discussions that take place at the end of a lesson, after students have had the opportunity to explore a task in small groups, provide a forum in which students can share and discuss ideas and in which you can ensure that the key mathematical goals that you meant to target during the lesson are realized. When engaging in reasoning-and-proving tasks, "classroom presentations of different arguments can furnish a forum for discussing with students the question of what constitutes proof—a notion for which many students have an inadequate understanding" (Knuth, 2002a, p. 489).

In orchestrating a productive mathematical discussion, you must determine which solutions to share and in what sequence, how to make connections between different solutions and to the key ideas of the lesson, and how to hold students accountable for participating in and actively listening to the discussion (Smith & Stein, 2018). Imagine that the students in your class produced the solutions to the

Building a Staircase task shown in Figure 4.3. What could you do to facilitate meaningful mathematical discourse during the whole-group debriefing of the solutions?

You may want each of the four students to present their solutions, since they represent different ways of thinking about the problem and can be used to focus on the mathematics that you targeted in your learning goals for the lesson. You might choose to sequence the solutions in the following order—B, C, A, and D—which goes from the most concrete (B) to the most abstract (D) solution. This sequencing, moving from visual approaches to an algebraic approach, would allow your students to access the discussion initially and ultimately consider the way in which the visual approach in Solution A is related to the algebraic approach in Solution D.

It is also reasonable to consider sharing an approach that is not yet complete and having the class discuss together the next steps or having a student or group that carried the same strategy through to completion discuss how they finished the task. For example, one of your students may notice the triangular nature of the staircase (see Figure 4.9) and may know the formula for finding the area of a triangle ($A = \frac{1}{2}bh$), but not be sure how to account for the jagged edge of the triangle. This student could present his or her incomplete solution to the class and ask for help from classmates in finishing the argument. Or, you could ask the student to present the incomplete solution to the class and ask groups to work on completing this solution path.

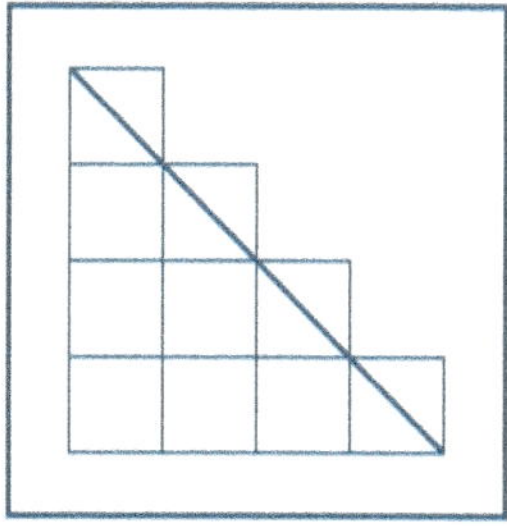

In addition to having each student explain his or her strategy and solution to the class, it is important for you to ask questions to the entire class that make the mathematics you are targeting salient and to help students in making connections between the different solution strategies. Because the goal of using this task is to build students' capacities for reasoning-and-proving, perhaps the most important question to ask about each solution is if the student's argument counts as a proof. This would help students evaluate each solution in light of the criteria for judging whether an argument counts as a proof (Figure 3.3) and, in so doing, strengthen their understanding of proof. In addition, you could pose a question about how Student B's proof (generic argument) is similar to or different from the three other proofs (which are demonstrations) or about how Student A's and Student D's solutions are related. Finally, you may also want students to discuss the three different algebraic generalizations $\left(\dfrac{n(n+1)}{2}; \dfrac{n^2+n}{2}; \dfrac{1}{2}n^2 + \dfrac{1}{2}n \right)$ and how and why they are equivalent. (See Lloyd, Herbel-Eisenmann, & Star, 2011, for an extended discussion about equivalent expressions.) These questions will help students move beyond their own solutions to the problem and to consider the mathematical ideas at the heart of the lesson.

Ultimately, it is important to ensure that debriefing "discussions" in your classroom are not just a sequence of students showing and telling, with all of the dialogue occurring between you and the presenting students. While you may ask the presenting student initial questions about a strategy and solution, you can help the other students in the class improve their listening skills and learn from each other's solutions by asking one or more of them to repeat what the presenting student said in their own words and/or to explain the strategy on a different concrete example (e.g., explain how the strategy presented in Student C's solution would be used on a five-step staircase). This helps to ensure that students understand the presented solutions. Students should also be encouraged to ask questions of each other.

We presented one scenario of how you could select, sequence, and discuss the student work from Figure 4.3. The way in which *you* decide to select, sequence, and discuss a set of student solutions depends on your learning goals—what you want students to learn about mathematics as a result of engaging in the lesson. Well-articulated learning goals will help you determine the mathematical ideas that you want to make salient in the lesson, the solutions that will help you make the targeted mathematics public, and the questions that will focus your students' thinking. We will return to the importance of establishing mathematics goals to focus learning in Chapter 5.

   WE REASON & WE PROVE FOR ALL MATHEMATICS

Pause and Consider

1. How might focusing on the four effective teaching practices (listed below) help you support students to develop their capacity to reason-and-prove?
   - Implement tasks that promote reasoning and problem solving
   - Use and connect mathematical representations
   - Pose purposeful questions
   - Facilitate meaningful mathematical discourse
2. How might focusing on the four effective teaching practices help you to address the challenges you identified at the beginning of the chapter?

Record your thoughts in your journal. If possible, discuss your responses with colleagues.

DETERMINING HOW STUDENT LEARNING IS SUPPORTED: THE CASE OF VICKY MANSFIELD

In Activity 4.1, you will read and analyze *Writing and Critiquing Proofs: The Case of Vicky Mansfield*. Vicky Mansfield is starting her third year of teaching mathematics at Knight High School. She uses the district-adopted, problems-based textbook that is intended to develop students' ability to reason about and make sense of the mathematics they are learning and ultimately to help them see proof as a natural mathematical activity that helps explain why things work. Prior to the lesson depicted in the case, Vicky's students completed a lesson where they learned about intersecting line theorems. In this lesson, Vicky wants her students to use that knowledge, along with previously learned definitions, theorems, and properties, to write a flowchart proof that proves a geometric conjecture.

Activity 4.1 Supporting Learning in Vicky Mansfield's Classroom

Read *Writing and Critiquing Proofs: The Case of Vicky Mansfield* that appears in Appendix C. Respond to the prompts on the next page in your journal. You can download a recording sheet to facilitate your work from the companion website.

continued>>

<<continued

- Identify ways in which Vicky supported her students' learning through her use of the four effective teaching practices:
  - Implement tasks that promote reasoning and problem solving
  - Use and connect mathematical representations
  - Pose purposeful questions
  - Facilitate meaningful discourse
- Consider whether or not the whole-group discussion was productive and provide evidence to support your perspective.

online resources — resources.corwin.com/reasonandprove

Download The Case of Vicky Mansfield recording sheet.

Debriefing Activity 4.1: Supporting Learning in Vicky Mansfield's Classroom

Vicky wanted her students to prove conjectures in different ways, drawing on their accumulated knowledge of geometry, and to critique the reasoning of others in order to provide more clarity and completeness to their arguments. Vicky's students created two different proofs and raised questions about the proofs that had been created, if at first somewhat reluctantly. The success of Vicky's lesson was due in large part to her use of the four effective teaching practices.

Implement tasks that promote reasoning and problem solving. Rather than giving students the task as it appeared in their book (Figure C.2 in Appendix C), Vicky began the lesson by presenting her students with a diagram (see Figure C.1 in Appendix C) and asking them to make as many observations as they could about the diagram. Vicky then asked students to read Parts A and B of the task and decide what conclusions they could come to about the angles in the diagram. She then asked students to make a conjecture about the relationships between lines AC and DE, and then create a flowchart proof that would prove their conjecture.

The task, as modified by Vicky, provided an opportunity for all students to make observations about the diagram. These initial observations then became a resource on which to draw when students moved on to subsequent parts of the task. The task was high-level since there

WE REASON & WE PROVE FOR ALL MATHEMATICS

was no pathway stated or implied, there were multiple entry points, and the task could be solved in different ways. The task had a "low floor"—everyone in the class was likely to notice something in the diagram that would be useful in other parts of the task—and a "high ceiling"—since creating a proof would require assembling a general argument with a logical arrangement of statements, a more challenging activity. The task was aligned with her goals for students' learning and provided students with the opportunity to make conjectures and develop arguments—important components of reasoning-and-proving.

Use and connect mathematical representations. The students in Vicky's class had been introduced to the flowchart format for writing proofs earlier in the week. This is a proof model that may not be commonly found in high school geometry books, and one that allows students to show their line of reasoning. In this particular lesson, students used a flowchart (which can be considered a type of visual representation) to organize a proof. Since in this lesson students were learning how to use a flowchart model, we did not see a range of different representations being utilized within their solutions, although the proof depended on the visual representation of lines AC and DE as well as the written text about angles DBA and CBD.

Pose purposeful questions. As Vicky interacted with different small groups, she asked questions to assess what the students understood and to advance them toward the lesson learning goals. For example, the students in Group 3 indicated that they were stuck. Vicky began her interactions with the group by asking students what they had noticed about the diagram (assessing). Students responded by saying that Angle 1 was equal to Angle 2. Vicky then asked students to go back to their observations and see if there was anything else they could say about Angles 1 and 2 (assessing). Students acknowledged that if Angles 1 and 2 were equal and their sum was 180 degrees (something several group members had noted), then Angles 1 and 2 had to be right angles. This finding led students to conclude that lines AC and DE must be perpendicular. At this point, Vicky indicated that they had made some interesting points but that they now had to figure out how to arrange their ideas into a flowchart that others could understand (advancing). Then she walked away.

Although the case only captured Vicky's interactions with two groups, in both episodes we see Vicky trying to make sense of what the students did, withholding judgment about the correctness of what they had done, and using what emerged from their discussion as the basis for moving them forward toward the lesson's learning goals. In both episodes, after she asked her assessing and advancing questions, she left the group to continue to collaborate and work toward completing a proof.

Facilitate meaningful discourse. Vicky built the discussion that took place in the last 30 minutes of her class on the work that students produced. Vicky asked Groups 1 and 5 to record their proofs on whiteboards that were placed at the front of the room, visible to the class. She decided to start the discussion with Group 1's proof since their proof was similar to the proofs created by most other groups. By starting with a proof that was familiar to many students, they would be better positioned to make sense of it and offer critique on the argument. Vicky decided to conclude the discussion by focusing on Group 5's proof, which was very different from what others had produced. By saving the "different" proof until later in the discussion, she hoped that students would be more willing to ask questions and try to make sense of other students' reasoning.

Rather than asking the authors of the proofs to explain their proofs to the class, Vicky asked the class "to be critical in a respectful way," raising questions about things they did not understand and places where they thought additional explanation was required. This put students in the position of trying to make sense of someone else's reasoning—not an easy thing to do. They did not respond immediately to Vicky's inquiry, but after she allowed wait time and offered another invitation to ask questions, students began to ask questions and suggest revisions.

There were key moments in the discussion of each proof when Vicky asked a question that helped students clarify and refine the proof. For example, in the discussion of Group 1's proof, Katie explained that before you can say that 2 times the measure of Angle 1 is equal to 180 degrees, you have to indicate that the measure of Angle 1 plus the measure of Angle 1 is equal to 180 degrees. At this point, Vicky asked, "Does she need to say *what* she substituted?" This question led Katie to add the fact that she was substituting the measure of Angle 1 for the measure of Angle 2. Towards the end of the discussion of Group 5's proof, Vicky asked whether the statements made on the top of the proof (see Figure C.9 in Appendix C) were sufficient to prove that all four of the angles were really the same. Again, this question led to a discussion among students that resulted in revising

　　　　　　　　　　　　　　　WE REASON & WE PROVE FOR ALL MATHEMATICS

the proof to include statements regarding substitution. While Vicky solicited students' contributions throughout the discussion, she was ready to ask a question that would push students to consider some aspect of the proof that she felt needed to be attended to.

At one point in the discussion, Adam raised a question, which led to what appeared to be an unplanned exploration. Adam's question, "But then how would we prove that the four angles add up to 360 degrees?" led to Stacie and Paige's conjecture that angles made by lines that intersect at a point will always add up to 360 degrees. While Vicky gave students time to explore the conjecture during class, she decided to have students continue to explore this conjecture over the weekend and returned to the discussion of Group 5's proof. In so doing, Vicky used a conjecture made by students as a target for further exploration, thus honoring the thinking of the students but still keeping her eye on the goals of the lesson.

Was the discussion productive? In judging whether or not the discussion was productive, it is important to view the discussion in light of what Vicky wanted students to learn during this lesson: (a) Definitions, postulates, and theorems can be used in proving conjectures; (b) conjectures can be proven in different ways; and (c) critiquing the reasoning of others can result in greater understanding, clarity, and completeness. Vicky appeared to be making progress on all three of her goals. Students used definitions, postulates, and theorems as needed in creating their proofs. The proofs offered by Groups 1 and 5 showed that there were different ways to prove a conjecture. The discussion around the two proofs made it clear that the comments provided by the students in the class added to the clarity and completeness of the proofs. While Vicky planned to continue the discussion of Group 5's proof on Monday, they had clearly made progress on what she had set out to do.

The success of the lesson was due in large measure to Vicky's advance planning and her careful attention to what students were doing and saying during the lesson. During her planning, Vicky made important decisions regarding the goals for the lesson and the nature of the task in which she would ask students to engage. The modifications she made to the task as it appeared in the book made it accessible to all students in the class. During the lesson, Vicky walked around the room making notes on her lesson plan. As a result, she knew what each group had done and was able to make informed decisions about which groups to have present and in what order. Her questions throughout the discussion served to invite student participation and, at times, focus students' attention on key aspects of the proofs under discussion.

Pause and Consider

Return to the page in your journal that you started in Chapter 2, titled "Pedagogical Moves That Support Students' Capacities to Reason-and-Prove." Reflect on the pedagogical moves that Vicky made during this lesson and add to your list.

DETERMINING HOW STUDENT LEARNING IS SUPPORTED: THE CASE OF NANCY EDWARDS

In Activity 4.2, you will read and analyze *Pressing Students to Prove It: The Case of Nancy Edwards*. At the time that she taught this lesson, Nancy Edwards was beginning her second year of teaching. She admitted that during her first year of teaching, she was in survival mode most of the time. She considered it to be a good class when she got through it with no major disruptions and kept up with the district pacing guide. Although she challenged students on occasion by engaging them in a problem-solving activity, she mostly stuck to the book and taught her students the procedures they needed to know and provided ample time for them to practice and gain proficiency in using those procedures. In this case, she implements a task that she was introduced to at a recent professional development session—one that requires students to engage in reasoning-and-proving.

Activity 4.2 Supporting Student Learning in Nancy Edwards's Classroom

Read *Pressing Students to Prove It: The Case of Nancy Edwards* that appears in Appendix D. Respond to the following prompts in your journal. You can download a recording sheet to facilitate your work from the companion website.

resources.corwin.com/reasonandprove

Download The Case of Nancy Edwards recording sheet.

- Identify ways in which Nancy supported her students' learning through her use of the four effective teaching practices:
    - Implement tasks that promote reasoning and problem solving
    - Use and connect mathematical representations
    - Pose purposeful questions
    - Facilitate meaningful discourse
- Consider whether or not the discussion was productive and provide evidence to support your perspective.

Debriefing Activity 4.2: Supporting Student Learning in Nancy Edwards's Classroom

Nancy Edwards wanted her students to make sense of mathematics and not just learn to apply routine procedures and repeat memorized definitions and theorems. The lesson portrayed in the case suggests that Nancy's students were learning what constitutes a convincing argument and how to create one. The four effective teaching practices were key to supporting her students' learning and engagement.

Implement tasks that promote reasoning and problem solving. Nancy selected a task that was aligned with her goal for student learning. The task was high level—there was no prescribed pathway for solving the task, and students were encouraged to "use words, pictures, numbers, or anything else they could think of" in creating their arguments. The task can be characterized as "low floor high ceiling"—"low floor" because students could enter the task by trying examples (with numbers, pictures, or models) and looking for patterns and making conjectures, and "high ceiling" because creating a valid argument required looking for the underlying structure, a more challenging aspect of the task.

Use and connect mathematical representations. When Nancy first presented the task to the class, she encouraged students to use different representations in creating their arguments. In her initial visit to Group 6, she suggested that students try to "build a model or draw a picture" to show that when you divide by two, "there is one left over."

She encouraged Groups 1 and 2 to find a way to connect their solutions, both of which argued that every odd number has one left over and that the two "leftover ones" form an even number, but did so using different representations. During the whole-group discussion at the end of class, Yolanda (Group 1) and Jamilla (Group 2) explained that the "extra 1 left over" in Group 1's picture was the same as the "loner number" in Group 2's logical argument.

At Nancy's request, Ramon (Group 5) explained that his group's statement that an odd number was an even number + 1 was the same as Group 2's statement that "every odd number has one loner number" and Group 1's picture that showed one leftover. When it was Ben's (Group 3) turn to describe what his group had done, he claimed, "Ours isn't very different, we just used algebra from the beginning." He went on to explain that they had used $2x$ and $2a$ to represent "EVEN." The different representations used in the class served to show that there were different ways to think about and represent the situation, but that they could all be connected to each other.

Pose purposeful questions. Nancy's interaction with each group followed a similar pattern: ask a "hip pocket" assessing question to determine how students are making sense of the problem and what they had produced so far, ask an advancing question that challenged students to think more generally, and leave the group with something new to pursue. It is worth noting that while the pattern of interaction was the same in her visit to each group, Nancy's specific suggestions and advancing questions were based squarely on the work that students had produced and the way they were already thinking about the task. As a result, students were able to move forward *based on their own thinking*.

For example, in her interaction with Group 1, Nancy asked the group to explain what they were doing (assessing). This gave the group the opportunity to explain the picture they had drawn and what it represented. Nancy, realizing that the group had shown that it worked for one case, asked whether this would convince a skeptic (advancing). This advancing question was intended to press the students to think beyond a single example. While the group realized they had to do more than just draw additional pictures, providing Group 2 as an additional resource allowed the group to see the generality in their picture. Leaving Groups 1 and 2 to "sort things out" without hovering over them sent the message that she believed they could do it.

Facilitate meaningful discourse. Nancy built the discussion that took place at the end of class on the work that students produced, beginning with a discussion of Group 1's picture and culminating with Group 3's algebraic solution. While Nancy wanted all students to understand the

    WE REASON & WE PROVE FOR ALL MATHEMATICS

algebraic approach, she decided to discuss it last since most groups had not used algebra. By starting with Group 1's picture and moving to increasingly more algebraic approaches, Nancy helped students build connections between different representations and see how algebraic symbols can be used to generalize the mathematical relationship.

While Nancy invited specific students to present the work their groups had produced, this was not a series of show-and-tell episodes. Nancy encouraged the students in the class to ask questions of the presenter. For example, Mike indicated that he was not sure what a loner number was, and Jamilla spontaneously provided a more detailed explanation. Osmi asked Ben how he knew to use $2x$ to represent an even number, and after Ben indicated that he remembered it from algebra, Kim jumped in with a more conceptual explanation regarding why this made sense. In addition, Yolanda and Jamilla joined the conversation about the posters produced by Groups 1 and 2 without being specifically invited to do so.

Was the discussion productive? As we said previously, in judging whether or not the discussion was productive, it is important to view the discussion in light of what the teacher wanted students to learn in the lesson:

1. For an argument to be a proof, it must show that it is true for all cases.

2. Proofs can utilize different representations, but they can be connected.

3. Examples alone do not constitute a proof.

Based on the evidence presented in the case, it is clear that Nancy was actively working toward these goals. Several different representations were presented and connected (Goal 2). While Group 1's poster featured a single example, the discussion of this solution made clear that the students had come to realize (through their collaboration with Group 2) that what they had found in their one example applied to all odd numbers. Students' responses to the third question of the homework assignment—Does the argument presented on Poster 6 count as a proof?—will give Nancy a clear sense of whether students understand the limitation of examples (Goal 3) and a starting point for discussion about the characteristics of proof and the necessary components of an argument that is considered a proof (Goal 1).

The success of the discussion was due to the work Nancy did both prior to and during the class. When students began work on the task,

Nancy walked around the class, making note of the strategies that different groups were using and the things she wanted to be sure to bring out in the whole-class discussion. She reviewed her notes prior to the whole-class discussion in order to determine which solutions she would have presented and the order in which they would be presented so that connections could be made across solution strategies and so that more concrete approaches were presented before more abstract (algebraic) approaches, thus allowing all students access to the discussion. She connected all of the solutions to each other and to key ideas of number theory (e.g., an even number is divisible by 2, an even number can be represented algebraically as $2n$ and pictorially as sets of two). In addition, Homework Question 3 focused students on considering what counts as a proof and the limitation of examples, a connection to a goal in the lesson.

While it is not specifically discussed within the case, it is reasonable to conclude that Nancy anticipated the approaches students were likely to take in solving the task as well as the misconceptions they might have and the questions she could ask. It is hard to imagine that she would have been able to support students as they worked on the task had she not thought through different solution paths prior to the lesson. Hence, Nancy engaged in what Smith and Stein (2018) have labeled *5 Practices for Orchestrating Productive Mathematics Discussion*. These practices—anticipating, monitoring, selecting, sequencing, and connecting—are intended to eliminate the amount of in-the-moment decision making during a lesson by focusing on advanced planning.

LOOKING ACROSS THE CASES OF VICKY MANSFIELD AND NANCY EDWARDS

As we noted previously, the purpose of Chapter 4 is for you to focus on how teachers can support the development of students' capacity to reason-and-prove with particular attention to examining mathematics classrooms through the lens of four effective teaching practices. In Activity 4.3, you will reflect on two questions intended to pull your work on *The Case of Vicky Mansfield* and *The Case of Nancy Edwards* together. Following that activity, you will revisit the list of challenges you created at the beginning of this chapter and consider how your work on the cases might serve to address some of these challenges.

Activity 4.3 Lessons Learned

What lessons have you learned from your analysis of the two cases—Vicky Mansfield and Nancy Edwards—that apply to teaching reasoning-and-proving more broadly? Return to the list of pedagogical moves that you started in your journal and add to it as needed. If possible, discuss the lessons you learned with colleagues.

Debriefing Activity 4.3: Lessons Learned

There are many things that can be learned from the Cases of Vicky Mansfield and Nancy Edwards. Perhaps first and foremost is the importance of planning a lesson in advance so that you are clear on what you are trying to accomplish mathematically and how students' work on the task can help you in reaching your lesson goals. Other "lessons learned" that you may have identified include the following:

- Task selection is critical to the success of a lesson, since the potential of the task sets the limit of what can be accomplished. It is interesting to see that Vicky modified a task from her adopted textbook and Nancy used a textbook task that had been modified. In both cases, the modified task increased students' opportunities to engage in reasoning-and-proving.
- Monitoring students as they work on a task is critical since it provides the teacher with information regarding how students are thinking about and making sense of the situation. Being aware of how students might solve a task, and the potential questions to ask, positions a teacher to interact in supportive ways with small groups.
- Selecting students to present their solutions, rather than asking for volunteers, provides a teacher with more control over the ideas that are made public for discussion. Through this process, the teacher can determine which solutions are most likely to help her accomplish the goals of the lesson and the order that is likely to give all students access to the discussion and highlight the key mathematical ideas that are to be learned. In addition, the selection of specific individuals to present group solutions gives the teacher the opportunity to

ensure that over time all students get a chance to demonstrate their competence in a public forum.

- Making mathematical connections among different solution strategies, and to the key ideas in the lesson, is critical in ensuring that the discussion is not just about confirming a correct answer and seeing that a task can be solved in different ways.

The key point to take away from the analysis you did on the cases of Vicky Mansfield and Nancy Edwards is not the specifics that occurred in either classroom. Rather, it is the fact that regardless of the context or the content of the lesson, there are generalizable pedagogical moves you can make that will improve students' opportunities to learn and engage in reasoning-and-proving.

Pause and Consider

What insights do you now have on how you might overcome some of the challenges you anticipated at the beginning of this chapter? Record your thoughts in your journal or notebook. If possible, discuss your thoughts with colleagues.

CONNECTING to Your CLASSROOM

After engaging in the activities in this chapter, it is now time for you to plan and implement a reasoning-and-proving task with your students. You could extend the work you began in Chapter 3 in which you identified a mathematical task that promotes reasoning-and-proving, or you could identify a different task (perhaps one of the tasks from this chapter) to use. Planning for implementing the task should be guided by considering how to support students' capacity to reason-and-prove through use of the four effective teaching practices discussed in this chapter. The goal is for you to begin to internalize these teaching practices, and you do that through planning and implementing instruction in which your students are required to reason-and-prove.

Both Vicky Mansfield and Nancy Edwards supported the development of their students' capacity to reason-and-prove through the *tasks* they posed, the *representations* they encouraged students to use and the ways they connected those representations, the *questions they posed,* and the *discussions they facilitated*. Take what you have learned about these four teaching practices and "try out" a reasoning-and-proving task with your students. If possible, work with a colleague or colleagues in planning and reflecting on the implementation.

Part 1: Planning

First, choose a "low floor high ceiling" reasoning-and-proving task and solve it in multiple ways, using multiple solution paths and representations. Try to develop both demonstration proofs and generic arguments as appropriate. Anticipating students' responses is important in planning for where they may get stuck and for writing assessing/advancing questions prior to implementation.

Considering the task you selected, continue your planning by reflecting on the following questions:

1. Vicky Mansfield and Nancy Edwards posed *tasks* that were challenging yet accessible to their students and provided students with opportunities to reason-and-prove. To what extent will the task you identified build on students' prior knowledge? To what extent will your task challenge students? What changes (if any) would you like to make to your task, and why?

2. What *representations* will support students' work on the task? What *representations* might you need to introduce to students either prior to or during their work on this task?

3. One central aspect of instruction in both Vicky Mansfield's and Nancy Edwards's classrooms was the nature of the questions they asked students. Anticipate the ways that students are likely to approach and solve the task and, for each strategy you come up with, generate questions you can ask to assess and advance students' learning.

4. Both Vicky Mansfield and Nancy Edwards were very purposeful in selecting and sequencing the solutions to be shared during the whole-class discussion. Review the solutions that you anticipate and consider which ones are most likely to help you achieve your

continued>>

<<continued

goals for the lesson and what sequence of solutions would allow all students access to the ideas you are targeting.

Part 2: Implementation

Teach the lesson you have planned in one of your classes. Audio or video record the lesson if possible and collect student work and other artifacts that were produced during the lesson. Using what you collected, reflect on the experience by considering the following questions:

1. Was the discussion productive? Why or why not?
2. In what ways did your planning prior to the lesson impact the outcome?
3. How did your use of the effective teaching practices help support students' learning and engagement?

*****

In this chapter, you began to think about how four of the eight teaching practices from *Principles to Actions: Ensuring Mathematical Success for All* (NCTM, 2014) can be viewed from a reasoning-and-proving perspective and how those practices can support students' capacities to reason-and-prove. Earlier in the chapter, we asked you to pause and consider why it is a "critical" pedagogical move to walk away from a student or group of students after asking an advancing question. As can be seen in the cases of Vicky Mansfield and Nancy Edwards, giving students time to work without hovering over them supports their confidence as mathematical learners and helps them to build perseverance. Just remember to cycle back to groups in a timely manner so that you can see what progress they have made on the question you left them to pursue!

We hope that from reading and analyzing the cases of Vicky Mansfield and Nancy Edwards, you have begun to develop a vision of what reasoning-and-proving can look like in secondary mathematics classrooms. Incorporating the effective teaching practices used by Vicky and Nancy into your teaching repertoire will not only support your students' capacities to reason-and-prove, but also support their mathematics learning overall. In Chapter 5, we turn to modifying textbook tasks to promote students' opportunities to engage in reasoning-and-proving.

1. In this chapter, we described four of the effective teaching practices: Implement tasks that promote reasoning and problem solving; use and connect mathematical representations; facilitate meaningful mathematical discourse; and pose purposeful questions. Which of these practices is currently a central feature of your classroom instruction? Which of these practices do you think will be most challenging for you to implement? Why? How can you get started in working on these practices?

2. The success of the lessons taught by Vicky Mansfield and Nancy Edwards was attributed in large measure to the fact that both carefully planned the lesson prior to instruction. What do you see as essential components of the planning process? How do these components contribute to lesson success?

3. In the section of this chapter about posing purposeful questions, we suggested that the use of "hip pocket" questions can help you begin mathematical conversations with your students. What are other "hip pocket" questions you could use in your classroom?

4. Revisit the Chapter 1 vignettes about the teachers at Hoover High School. How might more thorough lesson planning prior to the lesson, with attention to the effective teaching practices, have better positioned the teachers in supporting students during the lesson?

Modifying Tasks to Increase Their Reasoning-and-Proving Potential

Although we have argued throughout the first four chapters of this book that reasoning-and-proving transcends content and grade level, the reality is that most popular middle and high school textbook series limit the discussion of proof to geometry textbooks. This raises the question regarding how teachers can provide more opportunities to engage students in reasoning-and-proving when their textbooks don't support it. It is interesting to note that Andrew Wiles, who published the first valid proof of Fermat's Last Theorem in 1995 at the age of 45 (Fermat made his conjecture in the 1600s), found the theorem in a book in the library as a 10-year-old. He was instantly engaged based on the accessibility of the problem and dedicated much of his early career as a research mathematician to proving the conjecture that had stymied mathematicians for centuries.

Since it is unlikely that most students would find appropriate problems to work on without direction from the teacher, it is critical for teachers to adapt problems they currently have access to in order to improve their reasoning-and-proving potential. In this chapter, you will develop a set of strategies that can be used to modify or adapt curricular tasks in order to include more opportunities for your students to engage in reasoning-and-proving. In particular, in this chapter you will be asked to consider

- the importance of aligning a goal and task in order to maximize student learning outcomes;
- the extent to which the curriculum you use provides opportunities for students to engage in reasoning-and-proving; and
- a generalized set of strategies that can be used to modify existing tasks so that students have more opportunities to engage in the range of activities related to reasoning-and-proof.

HOW DO YOU MAKE TASKS REASONING-AND-PROVING WORTHY?

The cases of Vicky Mansfield and Nancy Edwards from Chapter 4 make evident that the rich mathematical tasks that they chose for their lessons had a great impact on the opportunities for students to reason-and-prove. As we discussed in Chapter 4, tasks provide students with opportunities to learn mathematics, and the choices that teachers make about the tasks in which they engage their students have an impact on what students come to understand about mathematics. We know from research that most teachers select the majority of their tasks from textbooks or online curricular resources (NCES, 2003; Stigler & Hiebert, 2004). We also know from research, however, that the majority of these textbook tasks do not provide students with rich opportunities to think and reason.

For example, Johnson, Thompson, and Senk (2010) used a reasoning-and-proving framework to analyze the proof-related opportunities in high school algebra and precalculus textbooks. Specifically, they coded lessons related to reasoning-and-proving as to whether they included the following elements: making a conjecture, investigating a conjecture, developing an argument, evaluating an argument, or other proof-related activities. Figure 5.1 contains the findings from this study.

These statistics paint a clear picture—very few tasks in high school textbooks include explicit attention to reasoning-and-proving activities. The authors of this study do suggest, however, that making changes in tasks, such as discussing properties and distinguishing between examples and proofs, can elevate these sorts of problems to a more worthwhile status.

A noted exception to this lack of tasks that promote reasoning-and-proving in textbooks is represented by textbooks or curriculum materials that can be described as "problem-based." *Problem-based curriculum materials* and textbooks typically contain mathematical tasks that require a high level of cognitive demand (Smith & Stein, 1998) and support a launch-explore-summarize instructional sequence. Students who learn from problem-based curriculum materials typically work in small groups to explore mathematical tasks and then engage in whole-group, teacher-led discussions of the important mathematics of the task (or series of tasks). These kinds of curriculum materials were first designed in the 1990s

Problem-Based Curriculum Materials: curriculum materials that contain tasks that require a high level of cognitive demand and support a launch-explore-summarize instructional sequence in which students typically work in small groups to explore mathematical tasks and then engage in whole-group, teacher-led discussions.

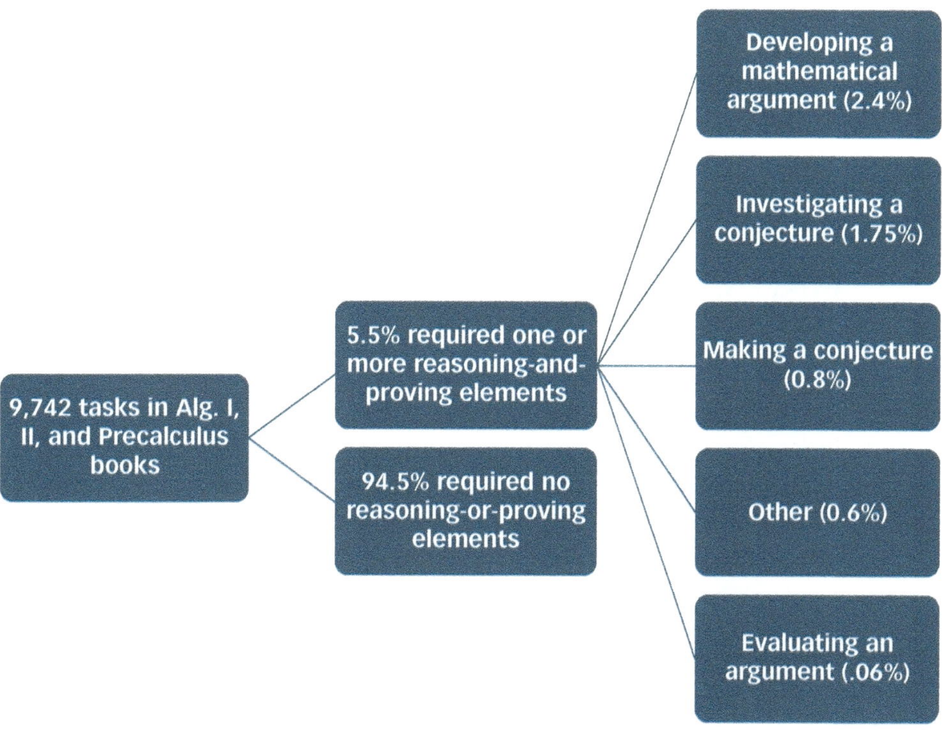

Source: Johnson, Thompson, and Senk (2010)

to embody the NCTM (1989) *Curriculum and Evaluation Standards for School Mathematics.* Although such curriculum materials exist, they are not adopted for use in middle and high schools with any consistency across the United States. Much more prevalent are the types of textbooks that Johnson et al. (2010) analyzed in their study.

G. Stylianides (2009) conducted a reasoning-and-proving analysis of one such set of middle school mathematics curriculum materials titled *Connected Mathematics Project* (Lappan et al., 1998/2004), where he analyzed 4,855 tasks across all of the units in three grades (6–8). While he found that 40% of those tasks engaged students in at least one aspect of reasoning-and-proving, he noted that the treatment of reasoning-and-proving was uneven across content areas, with number theory units providing the greatest opportunities for students to reason-and-prove.

This chapter is focused on helping you learn to adapt textbook tasks to enhance students' opportunities to reason-and-prove. We focus on task modification as an alternative to finding or inventing tasks that promote reasoning-and-proving, which can pose a greater burden on you as a teacher and may be more challenging to integrate into your existing curricular framework. Before we turn to modifying tasks, though, we present a discussion of the fifth of the focus teaching practice from *Principles to Actions: Ensuring Mathematical Success for All*—establish mathematics goals to focus learning.

RETURNING TO THE EFFECTIVE MATHEMATICS TEACHING PRACTICES

As we introduced in Chapter 1, *Principles to Actions: Ensuring Mathematical Success for All* (NCTM, 2014) identifies a set of eight research-based teaching practices that provide a framework for improving the quality of mathematics teaching and learning. We discussed how to *implement tasks that promote reasoning and problem solving* in detail in Chapter 4, which is a practice that serves as an important guidepost for the work of modifying mathematical tasks to strengthen their reasoning-and-proving potential. The teaching practice *establish mathematics goals to focus learning* also plays an important role in the work of this chapter.

Establish Mathematical Goals to Focus Learning

According to *Principles to Actions: Ensuring Mathematical Success for All* (NCTM, 2014), "The establishment of clear goals not only guides teachers' decision making during a lesson but also focuses students' attention on monitoring their own progress toward their intended learning outcomes" (p. 12). Establishing goals goes beyond citing the mathematics standard that a lesson and task address. Rather, goals should be connected to a curricular trajectory and to students' previous experiences. With respect to reasoning-and-proving, a goal to focus learning might take into account the types of reasoning-and-proving experiences students have had at a particular point in the year and steer students toward extending their reasoning-and-proving range (e.g., moving from forming a conjecture to making a proof argument) or deepening their fluency within a particular facet of reasoning-and-proving (e.g., using new representations in their argumentation or identifying patterns within a different mathematical topic).

Smith et al. (2017) make a distinction between learning goals and performance goals. ***Performance goals*** describe "a specific written or spoken performance that students should demonstrate as a result of the lesson" (p. 17). Performance goals can make the mathematical ideas that the teacher is interested in visible to students. These goals, however, are limited in that they describe a specific process and/or product—they describe what students can "do" as a result of the lesson. ***Learning goals*** provide targets for "what students can learn as a result of engaging in the lesson" (p. 17). These sorts of goals focus on the important mathematical understandings that a teacher wants students to develop in the lesson rather than describing the specific performances that may be evident in written or spoken work.

For example, consider these two lesson goals written for the implementation of the Building a Staircase task from Chapter 4 (Figure 4.2):

- Goal 1—Students will use multiple representations to write a proof.
- Goal 2—Students will understand how a visual representation of the staircase can be connected to the algebraic generalization and to the non-visual process of adding sums.

Goal 1 is a performance goal—it details what students will be able to *do* at the end of the lesson. Goal 2 is a learning goal—it details what students will *understand* by engaging in the activities in the lesson.

The work of designing and establishing goals goes hand in hand with identifying and modifying a mathematical task to enhance students' opportunities to reason-and-prove. Moreover, in modifying a mathematical task, it is important to consider the ways in which the task's goals might remain the same (e.g., you might want to ensure the same mathematics content is covered) and the ways in which the goals might be changed in the modified version (e.g., introducing new or different reasoning-and-proving processes). As we noted in Chapter 4, well-articulated learning goals will help you determine the mathematical ideas that you want to make salient in the lesson, the student solutions that will help you make the targeted ideas public, and the questions that will focus your students' thinking.

Performance Goals: goals that describe what process and/or product students will be able to "do" as a result of a lesson.

Learning Goals: goals that describe what students will understand mathematically as a result of a lesson.

Pause and Consider

Take a few moments to think about the goals that you establish for your mathematics lessons. You might want to consult your planning notes or the textbook you use. Do you establish performance goals or learning goals? If you establish mostly performance goals, choose a lesson and write a learning goal. Record both the performance goal and the learning goal in your journal and, if possible, discuss these with colleagues.

EXAMINING TEXTBOOKS OR CURRICULUM MATERIALS FOR REASONING-AND-PROVING OPPORTUNITIES

As discussed in the opening of this chapter, recent research has shown that many high school textbooks present few opportunities for students to engage in the reasoning-and-proving practices identified in the Reasoning-and-Proving Framework that we discussed in Chapter 3, particularly related to conjecturing and developing mathematical arguments (Johnson et al., 2010). In addition, many of the tasks that are labeled as opportunities for students to reason-and-prove are, in reality, very limited in the amount of reasoning that they ask students to provide. Activity 5.1 engages you in examining a mathematics textbook in order to understand the extent to which the tasks support students to reason-and-prove.

Activity 5.1 Analyzing Curriculum Materials for Reasoning-and-Proving Opportunities

1. Choose a textbook or resource that you are currently using or may use in the future to teach mathematics. If you do not have access to a mathematics textbook, you can access a website that has free curriculum materials (e.g., www.engageny.org/, eMathinstruction.com, or www.illustrativemathematics.org). If you and your colleagues have developed tasks and lessons, choose a unit from your collection.

2. Identify a chapter in the textbook or curriculum materials on which to focus this examination.

3. Focusing on the problem sets in that chapter (what the students will "do" to learn the mathematics), assess the extent to which students are required to engage in reasoning-and-proving while completing those problem sets. Refer to the Reasoning-and-Proving Framework from Chapter 3 in your analysis.

4. Complete the statement below in your journal. Be as specific as you can about the aspects of reasoning-and-proving that are supported by your textbook (e.g., does your textbook require that students look for patterns? Make and test conjectures? Provide non-proof arguments and/or proofs?) *My textbook provides the following kinds of opportunities for my students to engage in reasoning-and-proving:*

Debriefing Activity 5.1: Analyzing Curriculum Materials for Reasoning-and-Proving Opportunities

Dan Meyer, in a TED talk titled "Math Class Needs a Makeover" (http://www.ted.com/talks/dan_meyer_math_curriculum_makeover?language=en), notes that many textbook tasks that purport to be problem-solving tasks actually lead students through a set of procedures without allowing them the opportunities to think, reason, and make their own mathematical choices. During this activity, if you analyzed what we would describe as a traditional mathematics textbook, one that supports a "teacher demonstrates then students practice" instructional sequence, you may have come to a similar conclusion about the extent to which the tasks that students do have little potential to engage them in reasoning-and-proving. Meyer argues that typical textbook tasks limit student learning, and that all students need the opportunity to engage in mathematically meaningful tasks. He suggests that tasks in traditional textbooks can be easily modified in a number of ways to become worthwhile problem-solving opportunities. We argue the same when it comes to reasoning-and-proving.

You may want to deepen your analysis of the tasks you analyzed in Activity 5.1 by carefully considering the nature of the four aspects of reasoning-and-proving. Were students asked to make a proof argument, or would a non-proof argument address the problem well? Were there meaningful opportunities to make a conjecture? Was conjecturing authentic—did it arise from investigating patterns, for example, or was it based on a mathematical idea that students were already making use of and they were asked to conjecture about? How are words like *justify* and *explain* used?

If you analyzed a problems-based textbook, you may have noticed multiple opportunities for students to engage in reasoning-and-proving, similar to what G. Stylianides (2009) found in his textbook study. If so, as you move forward, it will be important for you to note which types of reasoning-and-proving activities are most prevalent in the textbook so that it will be clear what components are missing. This will help you consider how to implement the task modification strategies that you will generate as you engage in subsequent activities in this chapter.

Pause and Consider

Based on your experience with Activity 5.1, write a justification for why secondary mathematics teachers need to modify textbook tasks in order to provide more opportunities for students to engage in reasoning-and-proving activities. Discuss your justification statement with colleagues if possible.

REVISITING *THE CASE OF NANCY EDWARDS*

In *The Case of Nancy Edwards* (Chapter 4), Nancy asked her students to engage in making conjectures and constructing arguments related to the sum of two odd numbers. You may recall that Nancy used a modified version of the Sum of Two Odds task she had received at a recent professional development session. In Activity 5.2, you will reflect on Nancy's choice of task and consider both the original version of the textbook task and the modification. They are shown in Figure 5.2, with the original textbook version labeled as Task A1 and the modification as Task A2. (We use that labeling throughout this chapter for all of the tasks and their modifications.)

TASK A1	TASK A2
MAKING CONJECTURES—Complete the conjecture based on the pattern you observe in the specific cases.	Complete the conjectures below based on the pattern you observe in the examples. Then explain why the conjecture is always true *or* show a case in which it is not true.
Conjecture: The sum of any two odd numbers is _______.	**Conjecture:** The sum of any two odd numbers is _______.
$1 + 1 = 2$ $7 + 13 = 20$ $1 + 5 = 6$ $15 + 19 = 34$ $3 + 5 = 8$ $201 + 305 = 506$	$1 + 1 = 2$ $7 + 13 = 20$ $1 + 5 = 6$ $15 + 19 = 34$ $3 + 5 = 8$ $201 + 305 = 506$
Conjecture: The product of any two odd numbers is ______.	**Conjecture:** The product of any two odd numbers is ______.
$1 \times 1 = 1$ $7 \times 13 = 91$ $1 \times 5 = 5$ $15 \times 19 = 285$ $3 \times 5 = 15$ $201 \times 305 = 61,305$	$1 \times 1 = 1$ $7 \times 13 = 91$ $1 \times 5 = 5$ $15 \times 19 = 285$ $3 \times 5 = 15$ $201 \times 305 = 61,305$

You will work with a number of tasks and their modified versions throughout this chapter. To support your work on the chapter activities, we have created a document that contains all of the original tasks and their modifications that you can download from the companion website.

resources.corwin.com/reasonandprove

Download the Comparing Tasks and Their Modifications document to use as a reference throughout this chapter.

Activity 5.2 Comparing Tasks A1 and A2

Consider Task A1 and Task A2 in Figure 5.2. Record your responses to each of the three questions in your journal. If you are working on these activities with colleagues, compare and discuss your responses as a group before moving on.

1. How are Task A1 and Task A2 similar? How are they different?

2. What reason did Nancy Edwards give for using Task A2 instead of Task A1?

continued>>

<<continued

3. What opportunities for reasoning-and-proving does Task A2 provide that Task A1 does not? In what ways did Nancy Edwards' students appear to benefit from these opportunities?

Debriefing Activity 5.2: Comparing Tasks A1 and A2

In your comparison of the two tasks, you might have noticed some of the similarities and differences listed below. Read these similarities and differences, and compare them to what you noted in response to this activity. Make adjustments to your notes as needed.

Similarities Between Tasks A1 and A2

- Both ask students to complete a conjecture based on patterns given.
- Both use the same examples.
- The formatting and wording of both tasks are very similar.

Differences Between Tasks A1 and A2

- A2 requires that students explain their reasoning and/or generalize.
- A2 requires students to represent odd numbers in a generalized way.
- A2 requires that students show the relationship holds for every case.
- The wording in A2 promotes doubt—that the pattern students see may not always hold.

If you recall, Nancy's concern about the original Task A1 centered on two points. First, the task as written did not provide students with opportunities to engage in and share their thinking and reasoning. As such, the task does not fully connect to the reasoning-and-proving activity of making conjectures since the basis for the conjecture would not be clear in the original task. Did students arrive at their answer through thinking and reasoning, or did they simply guess? Second, Nancy noted that because of the six examples provided and no request to explain thinking, students could walk away from the original task with the impression that a set of six examples was sufficient to prove a conjecture. Both of these concerns informed Nancy Edwards' decision to use the modified version of the task.

CONTINUING TO EXAMINE TASKS AND THEIR MODIFICATIONS

In Activities 5.3 and 5.4, you will work with two different versions of the Construction Conjectures task. In Activity 5.3, you will engage in the modified version of the task (B2) and then consider the opportunities to reason-and-prove that you experienced through your work on the task. Then, in Activity 5.4, we provide the original version of the task (Task B1) and ask you to consider the ways in which we modified it to create Task B2. Through this work, you will begin to learn about small changes, or tweaks, that can be made to a task that can make big differences in learners' opportunities to reason-and-prove. You can use a straightedge and compass or a ruler to engage in Parts a-d. If you are working through this book with colleagues, discuss your work.

> **Teaching Takeaway:**
>
> You do not have to make major changes to tasks to enhance students' opportunities to reason-and-prove. Sometimes, a small tweak will make a big difference.

Activity 5.3 The Construction Conjectures Task (Task B2)

Use a ruler and the three steps outlined below to complete Parts a-d.

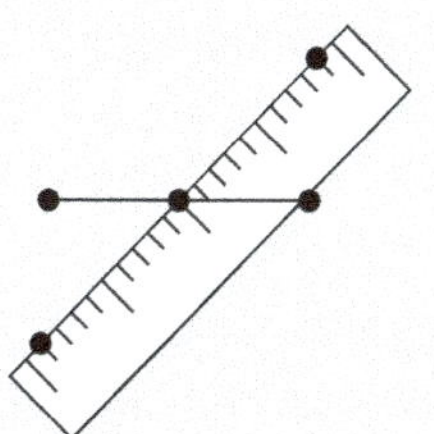

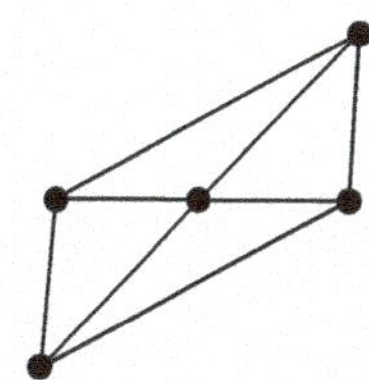

1. Use a straightedge to draw a segment. Identify and mark the midpoint.

2. Draw a second segment whose midpoint coincides with the first segment's midpoint.

3. Connect the endpoints of the two segments.

Record your work on the following questions in your notebook or binder.

a. Use this construction with a variety of starting segments.

b. Write a conjecture about the type of figure your construction produces.

c. Write a proof that explains why that figure is produced each time by the construction.

continued>>

<<continued

d. Revisit the "Criteria for Judging Whether an Argument Is a Proof" list (Figure 3.3) and assess the extent to which your proof meets the criteria. Modify your proof if needed.

Source: Adapted from Larson, R. et al., 2004.

Pause and Consider

What opportunities for reasoning-and-proving are afforded by this task? Record your thoughts in your journal or notebook. Discuss your response with colleagues if possible.

Debriefing Activity 5.3: Solving the Construction Conjectures Task

As you worked on the Construction Conjectures task, it is likely you considered some or all of the following mathematical ideas:

- You may have explored the ways in which the starting segment changed the figure you produced.
- You may have attempted to produce more specialized parallelograms. For example, if you used congruent segments and those segments bisected each other, then you created a rectangle, which could have also qualified as a square. If the segments were different lengths and were perpendicular bisectors of each other, you created a rhombus.
- The examples you chose likely influenced the conjecture you made. If you are working through this book with others, was it the same conjecture that your colleagues made? Are there multiple valid conjectures you could make related to this task? What prior geometry knowledge did you need to be able to justify your conjecture?

This task offers a number of opportunities for reasoning-and-proving. Students are likely to *identify patterns* based on creating a number of examples with different starting segments and noticing similarities and differences between them. Students are required to *make a conjecture* in Part b of the task, and to *generate a proof* in Part c. In Part d, they are asked to critically evaluate their proof by comparing it to the Criteria for Judging Whether an Argument Is a Proof (Figure 3.3) and making revisions if necessary.

A proof that justifies a conjecture can take a variety of forms, from a two-column proof to a flowchart proof to a narrative proof. Narrative proofs in this case are more likely to explicitly connect aspects of the construction to geometric principles. A two-column proof can make use of the same geometric principles, but the lack of explanatory text around it may leave the work of connecting to the three construction steps up to the reader. Figure 5.3 contains a sample conjecture and narrative proof for the Construction Conjectures task.

FIGURE 5.3 A narrative proof of a conjecture from the Construction Conjectures task.

Conjecture: The construction always creates a parallelogram.

Proof:

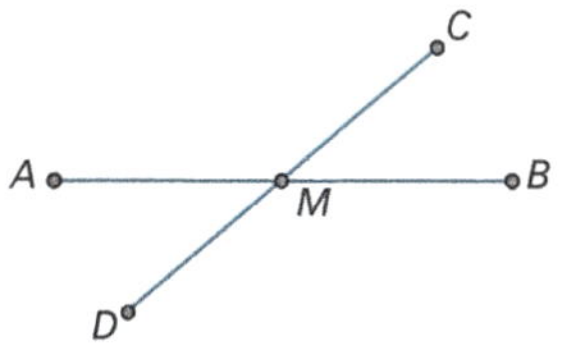

Constructing the segments as described in Steps 1 and 2 results in two pairs of vertical angles and two pairs of congruent segments:

$\angle AMD \cong \angle BMC$

$\angle DMB \cong \angle CMA$

$\overline{AM} \cong \overline{MB}$

$\overline{CM} \cong \overline{MD}$

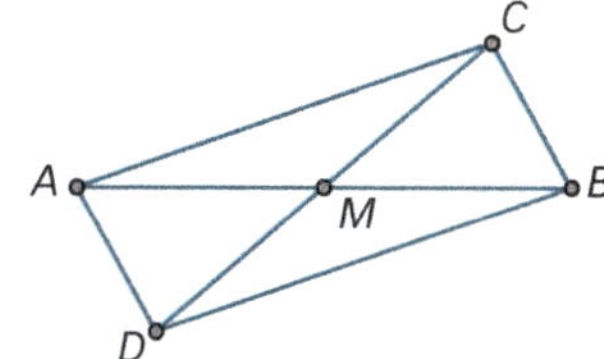

Connecting the segment endpoints and using the Side-Angle-Side postulate, we can identify two pairs of congruent triangles. We can identify pairs of congruent angles because corresponding parts of congruent triangles are congruent:

$\triangle AMD \cong \triangle BMC$

$\triangle DMB \cong \triangle CMA$

$\angle ADM \cong \angle BCM$

$\angle ACM \cong \angle BDM$

continued>>

FIGURE 5.3 (continued)

If a transversal intersects two lines so that alternate interior angles are congruent, the lines are parallel. Considering CD as a transversal intersecting AD and BC, since angles ADM and BCM are congruent, AD is parallel to BC. Similarly, considering AB as a transversal intersecting AC and BD, since angles ACM and BDM are congruent, AC and BD are parallel. Quadrilateral ACBD thus has two pairs of parallel sides, and is by definition a parallelogram. QED

In Activity 5.4, you will turn your attention to Task B1, the original version of the Construction Conjectures task you just completed (Task B2) as it appeared in a mathematics textbook, and then compare the original task and the modified version.

Activity 5.4 Comparing Tasks B1 and B2

Figure 5.4 contains the original version of the Construction Conjectures task as it appeared in a geometry book. Read this original task (Task B1) and compare it to the modified version of the task from Activity 5.3 (Task B2).

Task B1

REASONING WITH VISUALS: Explain why this method for drawing a parallelogram works. Name a theorem that supports your answer.

FIGURE 5.4 The Construction Conjectures task as it appeared in a textbook.

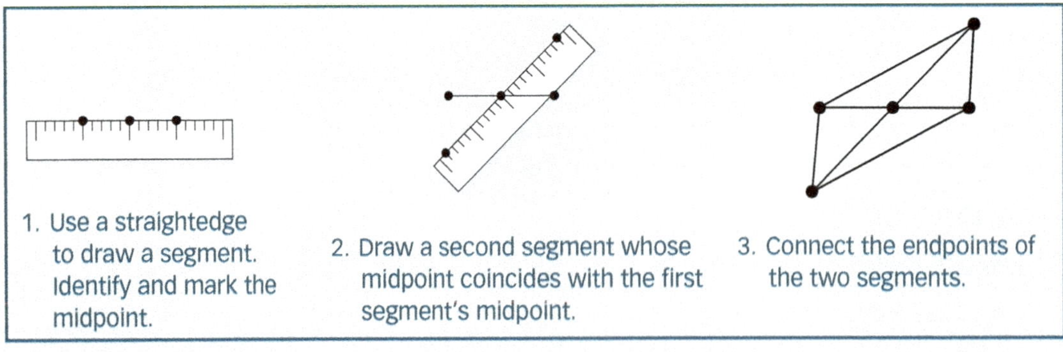

Source: Larson, R. et al., 2004

Record your answers to these two questions in your journal or notebook:

1. How is this original task (Task B1 in Figure 5.4) similar to and different from the modified version of the task that you did in Activity 5.3 (Task B2)?

2. In what ways did the modifications we made to Task B1 to create Task B2 expand students' opportunities to reason-and-prove?

Debriefing Activity 5.4: Comparing Tasks B1 and B2

Comparing Tasks B1 and B2 provides opportunities to consider the characteristics of a mathematical task that embodies the effective teaching practice *implement tasks that promote reasoning and problem solving*. Both tasks present some opportunities for reasoning and problem solving, but Task B2 provides stronger opportunities that are more closely connected to facets of reasoning-and-proving that we first presented in Chapter 3. We provide the following list of similarities and differences that you might have noticed in comparing the two tasks.

Similarities Between Tasks B1 and B2

- The constructions are the same.
- Both tasks deal with a general case.
- Both tasks require knowing/using the definition of a parallelogram.
- Both tasks require students to explain "why."
- Both tasks use visual representations.

Differences Between Tasks B1 and B2

- B2 requires that students seek patterns.
- B1 tells students that the construction makes a parallelogram.
- B2 requires conjecturing about the figure.
- B2 is hands-on—students have to make drawings.
- B1 asks for a theorem; B2 asks for an argument.
- B2 helps teachers identify misconceptions that students may have about parallelograms.

The two tasks both require that students consider the general case, but in Task B1, there is an implication that the case is true and the justification requested is a theorem (likely similar to *A quadrilateral is a parallelogram if and only if the diagonals bisect each other*). This implies that students have already been exposed to the theorem that is modeled in the task, which may make the task trivial for students who can immediately recall the theorem. If so, then no true conjecturing would be required in Task B1 because the theorem has already been presented. Identifying opportunities within a task to include more and/or different reasoning-and-proving activities is a promising direction for modifying textbook tasks.

REEXAMINING MODIFICATIONS MADE TO TASKS THROUGH A DIFFERENT LENS

You have compared Tasks A1 and B1 with their modifications (Tasks A2 and B2) and noted similarities and differences between the tasks as originally found in textbooks and modifications of those tasks. In Activity 5.5, you will return to those sets of tasks with a slightly different focus. In Activity 5.5, we ask you to articulate *how* Task A1 and Task B1 were modified to create Task A2 and Task B2.

Activity 5.5 Analyzing *How* Tasks A1 and B1 Were Modified

Reexamine Tasks A1 and A2 (Figure 5.2) and record *how* the original task was modified. You might begin with a statement like, "To improve the opportunities to reason-and-prove, Task A1 was modified in these ways …"

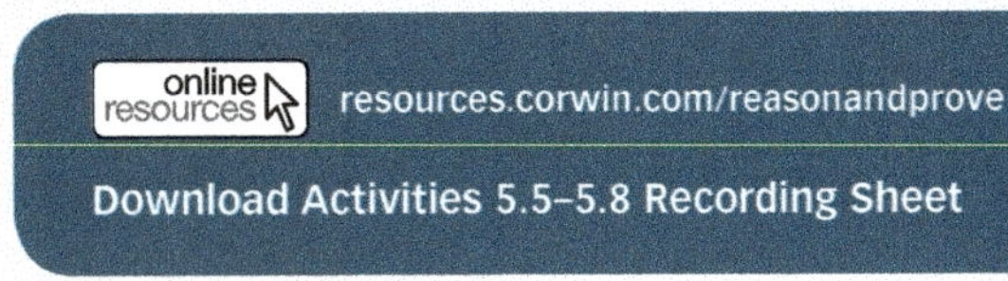

In a similar manner, reexamine Task B1 (Figure 5.4) and B2 (in Activity 5.3) and record *how* B1 was modified to create B2.

If you are working with colleagues, discuss your answers to this activity.

We have created a downloadable recording sheet (see Figure 5.5) that you can use to organize work in Activities 5.5–5.8.

   WE REASON & WE PROVE FOR ALL MATHEMATICS

	How the Original Task Was Modified *(in order to increase students' opportunities to engage in reasoning-and-proving)*
A1–A2	
B1–B2	
C1–C2	
D1–D2	
E1–E2	

Debriefing Activity 5.5: Analyzing How Tasks A1 and B1 Were Modified

In looking at *how* the original tasks were modified to enhance opportunities for reasoning-and-proving, you may have noticed several things. In Tasks A1 and A2, you might have noted that in A2, we added the direction for students to explain why their conjecture is always true, or to show a case in which it is not true. By asking students to explain their conjecture and by allowing for students to either prove or disprove the conjecture, the modified task (A2) strengthened the reasoning-and-proving practices of *identify a pattern* and *make a conjecture*, and added the opportunity to *develop a proof* or *develop a non-proof argument*.

For Tasks B1 and B2, you might have also noted a number of changes. We removed the statement in Task B1 that the construction creates a parallelogram, we added the direction to create more than one segment, and we added the prompt that requires students to conjecture and to explain why the construction always produces the same figure. These three modifications resulted in a task that promotes mathematical investigation and led to *identifying patterns*. Task B2 allows for authentic *conjecturing* by removing the prompt to name the theorem being used and opening up the means by which students can justify their statements. Task B2 also allows for explicit discussion of the relationships between empirical examples and proof, by asking specifically for a *proof argument* and connecting to the criteria for judging whether an argument counts as a proof. From a pedagogical perspective, because of the more open and exploratory nature of Task B2, the task has the potential to support you in learning what students know about the relationships between quadrilaterals and their diagonals, including opening up the opportunity for you to identify students' misconceptions about the mathematics.

COMPARING MORE TASKS WITH THEIR MODIFICATIONS

In Activities 5.6–5.8, we present three additional textbook tasks and their modified versions for you to examine. The goal for the remainder of this chapter is to generate a set of strategies, based on your analysis of tasks and their modifications, that you can use when modifying text-book tasks to improve students' opportunities to reason-and-prove.

Activity 5.6 Comparing Tasks C1 and C2

Figure 5.6 contains Task C1 (original) and Task C2 (modified). As you did in Activity 5.5, record the *ways* that the original task was modified to improve opportunities to reason-and-prove.

FIGURE 5.6 Tasks C1 and C2.

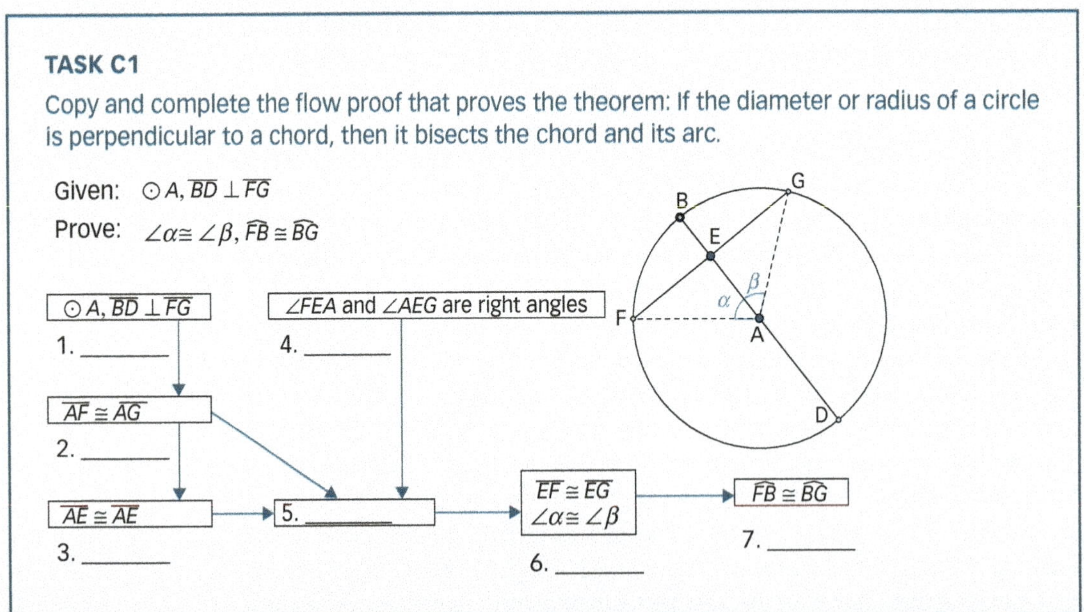

Debriefing Activity 5.6: Comparing Tasks C1 and C2

The first thing you likely noticed in Tasks C1 and C2 was that the skeletal version of the flowchart proof, with some statements filled in, is absent in C2. So, in C2 students are asked to make observations first and then to create their own proof. This modification did not technically change any of the reasoning-and-proving activities—one could argue that both tasks ask students to *generate a proof.* The modification, however, does substantially increase the thinking and reasoning demands for students in generating a proof. In Task C1, students only have to follow the proof structure and identify the reasons for the steps that the author took in writing the proof. Students must identify only one step on their own that is not named. In Task C2, students first make observations about the figure that are likely to help them *identify patterns* and use those patterns to consider how to *construct an argument.* They can also use those observations in any order that is appropriate to build their proof.

In Activity 5.7, you will compare Tasks D1 and D2 and record the *ways* that Task D1 was modified to create Task D2. Continue to record your responses about how the task was modified in your journal or on the recording chart you began using in Activity 5.5.

Activity 5.7 Comparing Tasks D1 and D2

Figure 5.7 contains Task D1 (original) and Task D2 (modified). Examine the two tasks and record the ways that D1 was modified to improve opportunities to reason-and-prove.

FIGURE 5.7 Tasks D1 and D2.

TASK D1

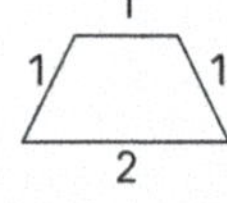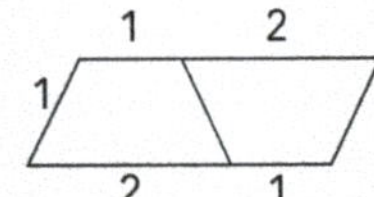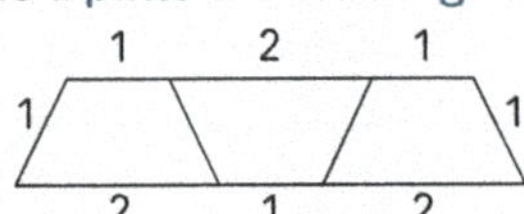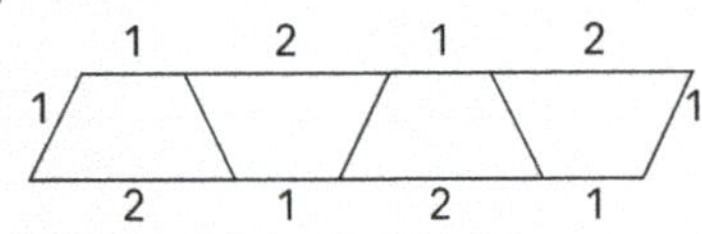

Perimeter = 5 units   Perimeter = 8 units   Perimeter = 11 units   Perimeter = 14 units

1. Create a formula that finds the perimeter of the trapezoid pattern for *n* trapezoids.
2. What is the perimeter for 12 trapezoids?

TASK D2

Use the diagram below that shows a pattern consisting of trapezoids.

1. Make as many observations as you can about the trapezoid pattern.
2. Find the perimeter of the first four trapezoid patterns shown above.
3. Find the perimeter of the pattern that contains 12 trapezoids without drawing a picture.
4. Write a generalization that can be used to find the perimeter of a pattern containing any number of trapezoids.
5. Using words, numbers, and/or connections to the visual diagram, prove that the generalization you created in #4 will always work.

   WE REASON & WE PROVE FOR ALL MATHEMATICS

Tasks D1 and D2 both use the same basic diagram and share similar questions. Task D1 only requires learners to create a formula; the second question can be then solved by the application of that formula. Task D2 provides greater access for students' engagement in reasoning-and-proving by asking them, in #1, to make observations about the trapezoid pattern. Pedagogically, a teacher could pause the work after #1 and record students' observations in a public space, which would allow students to have access to a wide range of ideas to build upon for the remainder of the task. In addition, if the teacher tags each idea with the name of the contributor, it acknowledges individuals' contributions, which is important in helping students to build a positive mathematical identity.

Teaching Takeaway:

Starting a reasoning-and-proving task with a prompt that asks for observations creates a "low floor high ceiling" task. Collecting those observations on the board before moving on acknowledges students' contributions and also supports students' work.

Task D2 makes use of the second question in Task D1 as part of a sequence of questions (1–3) that encourage *identifying a pattern*. Students then have to write a generalization and prove that the generalization would always work, encompassing *generate a proof*. So, Task D1 was modified by adding additional questions that specifically promote reasoning-and-proving.

Activity 5.8 presents one more pair of tasks for you to analyze for the *ways* that Task E1 was modified to form Task E2.

Activity 5.8 Comparing Tasks E1 and E2

Figure 5.8 on the next page contains Task E1 (original) and Task E2 (modified). Examine the two tasks and record *how* Task E1 was modified to improve students' opportunities to reason-and-prove.

continued>>

FIGURE 5.8 Tasks E1 and E2.

Task E1

Using the rules of exponents for products of powers, multiply the following monomials.

1. $x^3 \cdot x^5$
2. $y^2 \cdot y^2$
3. $a^3 \cdot a^2 \cdot a^4$
4. $(m^2 n)(m^3 n^3)$
5. $(5x^2 y^4)(3xy^3)$

Task E2

Complete the following calculations (calculators allowed):

1. $2^3 \cdot 2^5$
2. 2^8
3. $5^2 \cdot 5^4$
4. 5^6
5. $3^3 \cdot 3^2 \cdot 3^2$
6. 3^7
7. $4^2 \cdot 4^3 \cdot 4^2$
8. 4^7
a. Compare problems and answers in #1 and #2, #3 and #4, #5 and #6, and #7 and #8. What patterns do you observe?
b. Based on those patterns, write a conjecture about a procedure for multiplying monomials.
c. Write an argument in which you show that your conjecture will hold true when multiplying any number of monomials.

Debriefing Activity 5.8: Comparing Tasks E1 and E2

Task E1 is a typical set of problems found in most algebra textbooks. In the majority of textbooks, students are presented with the rule of exponents for products of powers (to multiply two powers having the same base, you add the exponents); are provided with the expanded form explanation; and then asked to apply the rule to a series of monomial multiplication problems (as represented in Task E1). We modified Task E1 to make it more exploratory in Task E2 by asking students to *identify patterns, write conjectures,* and *provide an argument.* We removed the direction in E1 for them to apply a given rule and replaced it with a series of prompts to lead students through the reasoning-and-proving process. As opposed to asking students to follow a stated rule for multiplying monomials, Task E2 helps them to generate the rule and then justify why that rule will always work. Note that Task E2 does not ask them to expand the monomials before multiplying—that kind of exploration should arise as they engage in writing an argument for why the monomial multiplication rule works. This modification opens up the task for an exploration of *patterns, conjecturing* about a possible method, and then *providing a non-proof explanation* of why the method always works.

STRATEGIES FOR MODIFYING A TASK TO ENHANCE STUDENTS' OPPORTUNITIES TO REASON-AND-PROVE

The work you have done in Activities 5.5–5.8 generated *ways* that Tasks A1–E1 were modified to enhance students' opportunities to reason-and-prove. You might notice some commonalities across the modifications in this task set, and you might have started to consider more generalized statements about how to modify tasks to enhance students' opportunities to reason-and-prove.

Activity 5.9 Generating Task Modification Strategies

Consider your answers to Activities 5.5–5.8 along with the content of the debriefs of those activities.

1. If needed, return to your answers and modify them based on the debriefs.

2. Then, in your journal, write three to five generalized strategies that can be used to modify tasks to enhance students' opportunities to reason-and-prove. You can start the list with a stem like, "To modify a task to enhance students' opportunities to reason-and-prove, a teacher can …"

Debriefing Activity 5.9: Generating Task Modification Strategies

Numerous strategies exist for modifying tasks in order to enhance their potential to engage students in reasoning-and-proving activities. We provide five such strategies in Figure 5.9 as a starting point for considering how to modify tasks and refer back to one of the modified tasks that utilized this strategy. We do not claim that these are the only strategies to use. They simply represent some specific actions that you can take as you begin to modify tasks to enhance students' opportunities to reason-and-prove. If you are working through this book with colleagues, stop and engage in a discussion about these modification strategies and others that may have been generated from engaging in this chapter.

FIGURE 5.9 Strategies for modifying tasks to enhance students' opportunities to reason-and-prove.

To modify a task to enhance students' opportunities to reason-and-prove, a teacher can

- engage students in investigation and conjecture instead of just giving answers. (Tasks A2, B2, D2, and E2)
- provide all students with access to a task by first making observations about a situation before moving on to more focused work. (Tasks D2 and E2)
- require students to provide a mathematical argument, proof, or explanation. (Tasks A2, B2, C2, D2, and E2)
- take away unnecessary scaffolding. (Task D2)
- ask students to explore a situation by generating empirical examples and looking for patterns. (Tasks B2, D2, and E2)

While these strategies provide some guidance about how to modify tasks to enhance students' opportunities to reason-and-prove, choosing how to modify tasks often depends on the original task. Does it provide too much scaffolding? If so, then removing that scaffolding might support students to explore, identify patterns, and make conjectures. Does the task not provide enough scaffolding to encourage reasoning-and-proving? If so, then adding prompts may be a way to modify the task. As you engage in the work of modifying tasks, you will need to depend on your understanding of reasoning-and-proving to make professional decisions about how to create a modified version that enhances your students' opportunities to engage in reasoning-and-proving.

As you use these and other strategies to modify tasks for your students, it is important to keep the following points in mind:

- The mathematics of the modified task should stay close to the mathematics of the original task, so that the task still meets similar mathematics learning goals. If the modified task doesn't provide the opportunity for students to engage with the same content, the task may no longer fit in the original instructional sequence. Carefully articulating *learning goals* as we discussed earlier in this chapter is a way to make sure that you have not modified the mathematics content that the students are supposed to be learning by engaging in the modified task. It is likely that modified tasks will have more learning goals when compared to the original task—that is okay. When you enhance students' opportunities to reason-and-prove, it makes sense that they will be learning more from engaging in the modified task than in the original task.
- The modifications that you make should enhance students' access to the problem, allowing them more entry points into the problem—creating a "low floor high ceiling" version of the task.
- Modifications should also encourage students to persevere and reason as a part of the task; strategies such as asking students to make observations can help provide them with the raw material to continue working and building mathematical arguments.
- And finally, although some strategies are about adding prompts to tasks, try not to tell students too much—consider judiciously the amount of information that you believe students will need to be successful with the task.

CONNECTING to Your CLASSROOM

1. Using your own curricular resources and the modification strategies from Figure 5.9 (as well as other modification strategies that you may have developed), select a task that you will be teaching soon and modify that task to enhance its reasoning-and-proving potential. Remember to keep the mathematical learning goal consistent if the original task comes from a particular content sequence in your textbook. If you are working with other teachers, prepare a cover sheet like the one in Figure 5.10, attach it to the original version of the task and the modified version, and then share the task and its modifications with your colleagues for discussion. You can download this cover sheet from the companion website.

2. Implement the modified task with your students. During implementation, use assessing and advancing questions (from Chapter 4) to support your students to make progress with the task. Consult with the list in your journal, titled "Pedagogical Moves That Support Students' Capacities to Reason-and-Prove," that you started in your journal in Chapter 2 and have continued to add to, to help guide your implementation. If appropriate, choose and sequence student responses for the whole-class summary of the task.

3. If you are working with other teachers, collect artifacts from your implementation and, following the lead of the Hoover High School teachers from Chapter 1, write a one-page summary of something that you found interesting and would like to discuss.

Source of Original Task

Resource Name ____

Publisher ____

1. How is the modified task similar to and different from the original task?
2. In what way(s) does the modified task enhance students' opportunities to reason-and-prove?
3. What specific strategy or strategies did you use to make the modification?

Goals and Implementation

4. What are your mathematical goals for the task with your class? Be sure to describe the ways in which the goals connect to students' prior experiences.
5. Implement the task with your students and reflect on the experience. How did the modifications of the task provide opportunities for reasoning-and-proving? In what ways did your goals inform the instructional decisions you made during the lesson?

*****

Modifying textbook or curriculum materials tasks can be more complicated than it may seem, and we are not arguing that you have to modify every task to provide more opportunities for your students to reason-and-prove. Start examining tasks that are in your materials with a reasoning-and-proving eye. If possible, discuss tasks that you think could be modified with a colleague for a second opinion. As time passes, you will become better and better at spotting those tasks that are "ripe" for modification. And, as you begin to implement the strategies for modification that were developed in this chapter, it will get easier. The time you put into modifying tasks at the beginning will pay off because you will get progressively better at noticing tasks that could be modified and have a toolbox of ways to modify the tasks to enhance students' opportunities to reason-and-prove.

Pause and Consider

As we move into the final chapters of this book, we suggest that you pause and reflect on what you have learned about reasoning-and-proving from engaging with Chapters 1–5. Take some time to revisit each chapter and note in your journal what new understandings you have built.

Discussion Questions

1. Why is establishing a clear goal for student learning in a lesson so important?

2. In Activity 5.1, you analyzed the textbook or resource you are currently using and determined the opportunities for reasoning-and-proving that it afforded. To what extent have you engaged your students in the reasoning-and-proving activities that are available in your materials? If you have not used these activities in the past, how might the effective teaching practices discussed in Chapter 4 help you do so in the future?

3. In Figure 5.9, we identified a set of strategies that can be used to modify tasks in order to increase students' opportunities to engage in reasoning-and-proving. Should every task you use be modified in some way? How will you decide when and if a task should be altered?

   WE REASON & WE PROVE FOR ALL MATHEMATICS

Using Context to Engage in Reasoning-and-Proving

"Realistic mathematics education" (RME) is a Dutch approach to learning and teaching mathematics. In RME, *context problems*, which are defined as "problems of which the problem situation is experientially real to the student" (Gravemeijer & Doorman, 1999, p. 111), have a major role in mathematics classrooms. In most mathematics curriculum materials available in the United States, context problems appear at the end of a section's practice/homework problem set under headings such as "Application Problems." Historically known as "word problems" in these kinds of textbooks, these problems situate the mathematics of the section in contexts and students are expected to use the skills they have developed in the section to solve the problems. In RME, however, context problems are the start of mathematical learning—not something for students to complete after they have learned the formalized mathematics—and they "can function as anchoring points for the reinvention of mathematics by the students themselves" (p. 111–112).

Context Problems: problems of which the problem is experientially real to the student.

In Chapter 4, we presented five different ways to represent mathematical ideas—visuals, symbols, language, contexts, and models (see Figure 4.5)—and discussed the importance of making connections between and among them. In this chapter, you will draw on the knowledge and tools you learned in previous chapters to consider how context problems can be used to support students' abilities to reason-and-prove. In particular, in this chapter you will be asked to consider

- the ways in which context problems may give students more access to reasoning-and-proving by providing a concrete grounding for making sense of a situation;

- how different representations can support student sense-making; and
- how teachers' actions, often well-intentioned, can limit students' learning opportunities.

HOW DOES CONTEXT AFFECT REASONING-AND-PROVING?

The activities in this chapter are all built around one reasoning-and-proving task—the Sticky Gum problem, shown in Figure 6.1. This task is different from most of the tasks that you have been working on up to this point (with the exception of the Build a Staircase task in Chapter 4) in that it is embedded in a context. The Merriam-Webster dictionary defines *context* as "the set of circumstances or facts that surround a particular event, situation, etc." (www.merriam-webster.com). Many contexts used in mathematical problems are considered to be "real world," but problems can also be situated within a context of science fiction, fantasy, or other fictional circumstances. Of particular interest for your work in this chapter will be attention to the ways in which context helps students make sense of a situation, create algebraic generalizations, and argue that those generalizations hold true for all cases.

FIGURE 6.1 A Sticky Gum Problem

Sticky Gum Problem

Ms. Hernandez came across a gumball machine one day when she was out with her twins. Of course, the twins each wanted a gumball. What's more, they insisted on being given gumballs of the same color. She can see that there are only red and white gumballs in the machine. The gumballs were a penny each, and there would be no way to tell which color would come out next. Ms. Hernandez decides that she will keep putting in pennies until she gets two gumballs that are the same color.

1. Why is three cents the most she will have to spend to satisfy her twins?
2. The next day, Ms. Hernandez passes a gumball machine with red, white, and blue gumballs. How could Ms. Hernandez satisfy her twins with their need for the same color this time? That is, what is the most Ms. Hernandez might have to spend that day?
3. Here comes Mr. Hodges with his triplets past the gumball machine in Question 2. Of course, all three of his children want to have the same color gumball. What is the most he might have to spend?
4. Generalize this problem as much as you can. Vary the number of colors. What about different-size families? Prove your generalization to show that it always works for any number of children and any number of gumball colors.

Source: Fendel, Resek, Alper, & Frazer (1996).

As we discussed in Chapter 4, the use of contextual representations, along with visual, symbolic, oral and written language, and physical representations, supports students in sense-making. We argue that the use of context also supports students in reasoning-and-proving.

CONSIDERING OPPORTUNITIES FOR REASONING-AND-PROVING

In Activity 6.1, you begin your work on the Sticky Gum problem by carefully considering the opportunities for reasoning-and-proving and other learning afforded by the task.

Activity 6.1 Opportunities for Learning From the Sticky Gum Problem

Read the Sticky Gum problem shown in Figure 6.1. Refer to Figure 3.4 (Reasoning-and-Proving Framework), Figure 4.5 (mathematical representations), and the Chapter 4 text about tasks that promote reasoning and problem solving to respond to the following prompts. To facilitate your work on this activity, you can download a copy of Figure 3.4 and Figure 4.5 from the companion website.

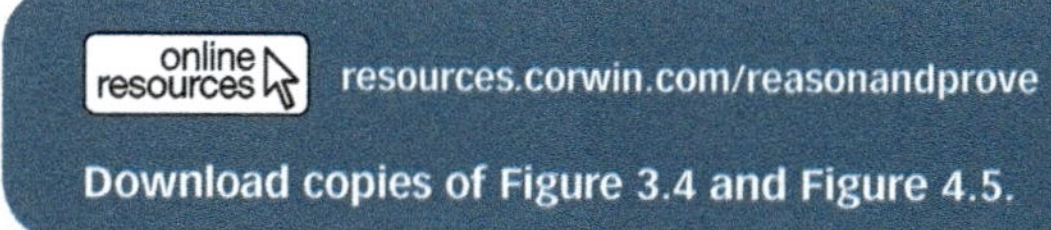

1. What opportunities for reasoning-and-proving are afforded by the task?

2. How does this task promote reasoning and problem solving?

3. What representations might be used in solving the task?

Record your responses in your journal. Discuss your responses with colleagues, if possible.

Debriefing Activity 6.1: Opportunities for Learning From the Sticky Gum Problem

The Sticky Gum problem promotes several components of reasoning-and-proving. First, students need to explore empirical examples, starting with three colors of gumballs and twins then triplets and ultimately trying different numbers of children and colors. These examples provide the foundation for identifying patterns and determining what is happening mathematically as the numbers of children and colors change. Then students are asked to make a generalization (conjecture about a mathematical relationship), and finally, students are asked to prove that generalization holds for any number of children and any number of gumball colors.

This task promotes reasoning and problem solving since there is no solution path stated or implied by the problem. There are multiple levels of entry to this task—students could start by modeling the context with a bag and colored tiles, drawing a picture to model what is happening in the context, or systematically recording results for different colors of gumballs and different numbers of children. The task also has the potential to engage students in challenging mathematics—most notably in proving a generalization. As discussed in Chapter 4, the task has a low floor and high ceiling.

The Sticky Gum problem also promotes making connections between a rich context (in this case, a real-world situation) and a ***mathematical generalization*** (a statement that defines the mathematical relationship between two or more entities for all cases and is most often presented with symbols or in narrative form) that represents the situation. Hence, the context here is more than a source of data to be abstracted and then abandoned—it can be utilized in making sense of the mathematical relationships.

This connection between the contextual and symbolic representations is an important connection for you to promote when engaging your students in reasoning-and-proving activities. Ellis (2007) cautions teachers against the common practice of always allowing students to develop symbolic generalizations solely from a set of examples or a table of values. She argues that *not requiring* students to also make connections to the contextual representation of a problem hinders their ability to construct an argument. While a table of values can be a supportive mathematical representation, it is important to help students relate the values within a table to the problem context in order to justify their generalization. A variety of representational forms (oral or written language, manipulative models, and pictures) can help students arrive at a generalization, construct an argument through making sense of the context, and critique the reasoning of others.

Refer back to the student work on the Building a Staircase task in Chapter 4 (Figure 4.2) as well as your own work on the Toothpick task in Chapter 3 (Activity 3.2). In what ways did the contexts of these two tasks facilitate reasoning-and-proving?

Record your thoughts in your journal and discuss them with colleagues, if possible.

SOLVING THE STICKY GUM PROBLEM

In Activity 6.2, you will solve the Sticky Gum problem and reflect on your use of representations. As you solve the Sticky Gum problem, revisit the activities that are part of the process of reasoning-and-proving: identifying patterns, making conjectures, providing non-proof arguments, and providing proofs. The context is intended to aid you not only with creating a generalization, but also in arguing why the generalization is always true. Hence, making connections among representations—the context, diagrams, and the written symbols—can serve as an aid in making sense of the mathematics and creating a proof.

Activity 6.2 Solving the Sticky Gum Problem

Solve the Sticky Gum problem (shown in Figure 6.1). In creating your argument (Part 4 of the task), be sure to make explicit connections between the context of the problem and any other representations you used.

If possible, discuss your work on the problem with one or more of your colleagues before moving on to the debriefing.

Debriefing Activity 6.2: Solving the Sticky Gum Problem

As you read through this debriefing of the Sticky Gum problem, you might find it helpful to make notes or draw a picture in your journal of the different scenarios that we present. If these ways of thinking about the problem are different from how you approached the problem, then making sense of the explanations that follow will benefit you in moving forward with the work in this chapter.

The most that Ms. Hernandez has to spend to satisfy her twins' request for matching gumballs is three cents because if the first two gumballs are different colors (one white and one red), then the third gumball is going to match one of the first two. So, the "worst-case scenario" for this context is if the first two gumballs are different colors. For twins and three gumball colors, the most Ms. Hernandez would have to spend is four cents. If the first three gumballs are all different colors (one white, one red, and one blue), then the fourth gumball will match one of the first three gumballs in color. Here, the worst-case scenario is defined when the first three gumballs are all different colors.

For three children and three gumball colors, the most Mr. Hodges would have to spend is seven cents. It is helpful to think about the way the gumball colors are drawn from the machine as either "different color sets" or "same color sets" when considering the worst-case scenario. For "different color sets," Mr. Hodges could draw one red, one white, and one blue and then again red, white, and blue. This would be a total of six cents, and the next gumball must be either red, white, or blue, so seven gumballs or a total of seven cents is the worst case. If you thought about the "same color set," then first two white gumballs might be drawn, then two blue gumballs and finally two red gumballs. Again, this is a total of six cents and the next gumball (the seventh one) will satisfy the triplets.

There are several equivalent forms of the generalization that you may have developed, such as

- $t = c(k - 1) + 1$
- $t = ck - c + 1$
- $t = ck - (c - 1)$

where t is the total number of cents spent, c is the number of gumball colors, and k is the total number of kids.

You may have used tables to explore relationships among the number of children, number of colors, and cost. You may have created a logical argument based on the context of the problem, or you may have used a picture or model to explain the phenomenon. The key here is that creating a generalization is a critical component in solving this problem, but it is not a proof. For proof, you need to explain why your generalization works.

Let's explore one of these generalizations: $t = c(k - 1) + 1$. This generalization holds no matter how many gumball colors or how many children, as long as the number of gumball colors is greater than or equal to the number of children. Let's consider the context again. Since the gumballs come out of the machine randomly, Ms. Hernandez will spend the most amount of money in the worst-case scenario. For Ms. Hernandez to get gumballs of the same color, the worst-case scenario is when one of each color is drawn (different color sets) before any duplicate colors are drawn. Thus, the worst-case scenario is dependent on the number of colors, not the total number of gumballs in the machine or the number of gumballs of each color. The number of sets, however, is dependent on the number of children (k). If there are two children, only one set of gumballs (containing one gumball of each color) would need to be drawn. The very next gumball is guaranteed to be a double of a color already drawn. If there are three children (remember that there would also need to be three or more gumball colors), two sets of gumballs (ensuring two gumballs of each color) would be drawn in the worst-case scenario. The very next gumball is guaranteed to be a triple of a color already drawn. The number of gumball colors (c) needs to be multiplied by one less than the number of children ($k - 1$), with one more gumball (+1) added to represent the final gumball needed to guarantee c number of gumballs that match.

This generalization also works for the second case of worst-case scenario, where the same gumball color is drawn consecutively, and changes colors just before there are enough gumballs for each of the children. Then the second color is drawn again and changes when a gumball is drawn for the last child. In this worst-case scenario, the sets of gumballs are all the same color, and the size of the set is one less than the total number of children ($k - 1$). The sets of same-color gumballs are drawn until the different gumball colors (c) expire. If this is the case, then for any number of children, k, and any number of gumball colors, c, where each kid must have the same gumball color, the total worst-case scenario cost at 1¢ per gumball is $c(k - 1) + 1$.

Assume the same color is drawn for each kid except one kid because if every kid got the same color, it would not be the worst possible case. Then a new color (second color) is drawn for every kid, but again it stops one short to match every kid. Then a third color is drawn from the machine, and again the number of gumballs of this third color is one less than the total number of kids. This situation is represented in Figure 6.2.

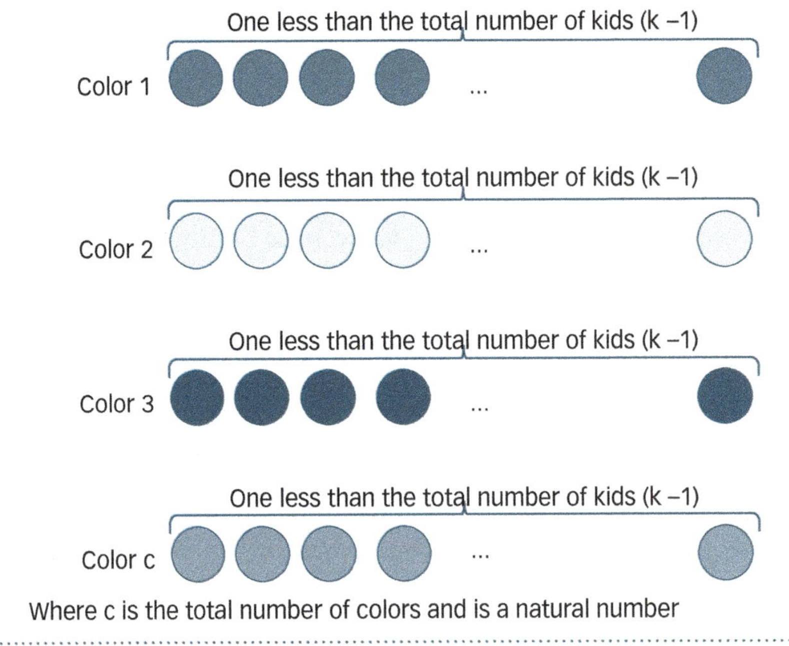

FIGURE 6.2 A pictorial representation of the Sticky Gum Problem.

So, after every gumball color, c, is drawn $k - 1$ times, then every kid has a matching colored gumball. And since there are new colors left in the machine, the very next gumball (+1) will be a match so that every kid has the same number of gumballs as the worst possible case.

Whew! That was a lot of hard mathematical thinking. Engaging in this task may have caused you to reconsider whether middle and high school students could engage in reasoning-and-proving through this task. As you will see from Activity 6.3, they can! In Activity 6.3, you will analyze a set of student work that was collected from high school classrooms where the Sticky Gum problem was implemented.

ANALYZING STUDENT WORK FROM THE STICKY GUM PROBLEM

Activity 6.3 engages you in ideas that we first presented in Chapters 2 and 3 about the limitations of empirical arguments as proof and the nature of reasoning-and-proving. One of the challenges of implementing tasks like the Sticky Gum problem is assessing students' abilities to engage in reasoning-and-proving by examining their written work. By using a combination of the Criteria for Judging Whether an Argument Counts as Proof (Figure 3.3) and the Reasoning-and-Proving Framework (Figure 3.4), you can create a rubric that supports your analysis of student work and assesses the extent to which students successfully engage in reasoning-and-proving. You may want to return to those chapters for a refresher before continuing with Activity 6.3. To facilitate your work on this activity, you can download a copy of Figure 3.3 and Figure 3.4 from the companion website.

Activity 6.3 Analyzing Student Work From the Sticky Gum Problem

Figure 6.4 contains eight pieces of student work (A–H) from the Sticky Gum problem. Using a chart like the one in Figure 6.3, analyze each piece of student work for evidence that the student engaged in reasoning-and-proving. Use the cells of the chart to make notes about the nature of the students' engagement in each component of reasoning-and-proving. For example, if you decide that a student has provided a non-proof argument, you can categorize it as an empirical argument or rationale or if you decide that a student has provided a proof, then you can categorize it as a demonstration proof or a generic argument. You can download a fillable version of Figure 6.3 from the companion website. Consult the Criteria for Judging Whether an Argument Counts as a Proof (Figure 3.3) as needed in your deliberations as well as the descriptions of non-proof arguments and proofs in Chapter 3. If possible, share and discuss your analysis with other teachers.

continued>>

<<continued

FIGURE 6.3 A chart to organize the analysis of Student Work A–H.

	Identifies Patterns	Makes a Conjecture	Provides a Non-Proof Argument	Provides a Proof
A				
B				
C				
D				
E				
F				
G				
H				

          WE REASON & WE PROVE FOR ALL MATHEMATICS

FIGURE 6.4 Student work samples from the Sticky Gum Problem.

Student A

many generalizations can come from these problems

children	colors	money
2	2	3
2	3	4
2	4	5
2	5	6
2	6	7

children	colors	money
3	2	5
3	3	7
3	4	9
3	5	11
3	6	13

children	colors	money
4	2	7
4	3	10
4	4	13
4	5	16
4	6	19
x	y	n

n=xy-(previous y)

multiply the number of children by the number of colors and subtract the previous amount of colors.

ex:

children	colors	money
4	2	7
4	3	10
4	4	13
4	5	16

4x4-3=13

continued>>

Student B

For problem number four, I drew in and out machines to see what kind of equations I could come up with.

For 2 children:

# of Candies	money
2	3
3	4
4	5
5	6

For 3 children:

# of Candies	money
2	5
3	7
4	9
5	11

Solution: 2. Ms. Hernandez would have to spend four cents on candy for her children.

3. Mr. Hodges would have to pay seven cents for candy for his children.

4. It can be said that to find out how much money you would have to pay with two children is just the number of the different colors of candy plus one ($x+1$, where x = the number of different colors of candy). To find out how much you would have to pay for three children, it is the number of different kinds of candy multiplied by two and added to one ($2x+1$, where x = the number of different colors of candy).

Here is the formula needed to rewrite problem 4 algebraically:

$$x = \text{colors}$$
$$y = \text{children}$$
$$z = \text{cents}$$

$$xy - (x-1) = z =$$
$$3 \cdot 3 - (3-1) =$$
$$9 - 2 = 7¢$$

The reason I chose this formula is as follows: I needed to multiply the colors by the children in order to get the maximum amount of money needed (including children getting more than one color of the same color). But since the children only have to have the same color as one of the gum balls, I needed to take away the other two possibilities, which is why I subtracted the colors minus 1. Look at the following diagram on the next page:

X: R W B

Y:

1¢	1¢	1¢
1¢	1¢	1¢
1¢	~~1¢~~	1¢

See, we don't need the last two results, of the triplets getting the same color of all the gum balls, just one color -- which is why we subtract the last two numbers, by taking the number of colors, and subtracting one, which in this case is 3-1, giving us 2, which ~~~~ we subtract from the kids times the colors, resulting in $3 \cdot 3 - (3-1) = 9 - 2 = 7$. Whew! Long sentence!

continued>>

Student D

Now, Mr. Hodges with his 3 Kids comes along to the 3
Color machine well I'm getting sick of writing out possibilities
I'm going to start on the real Pow now!
O.K. says my mind time for an in, out machine

K	C	A
2	2	3
2	3	4
3	3	7

K = # of kids   c = # of colors   a = how much $ you have to spend

After making this in, out machine I thought Wow each answer is 1
less than the K and C added that died as soon as I did Mr. Hodges
triplets but his was only one above so I spazzed out on the
original in, out machine I did and discovered that all the answers
were either 1 above or 1 less than K+C so after brainstorming
for awhile this is what I came up with for a final answer
$(K-1)C+1=A$ I then tried it with all the #'s I had in my
in, out machine and it worked Yeah!

4) Example for any amount of gumballs to match the color and amount

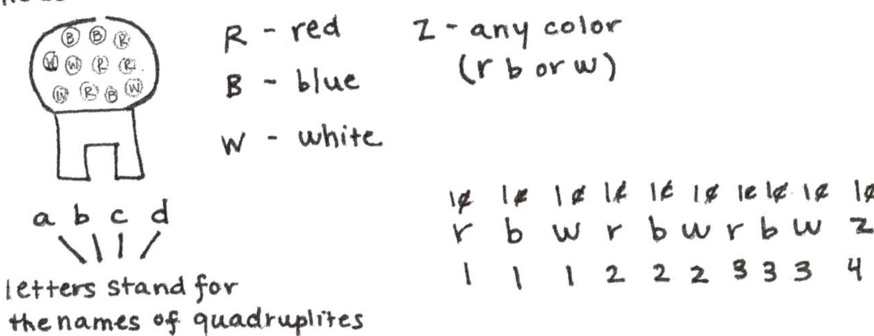

R - red

B - blue

W - white

Z - any color
(r b or w)

a b c d
\ | | /

letters stand for
the names of quadruplites

1¢ 1¢ 1¢ 1¢ 1¢ 1¢ 1¢ 1¢ 1¢ 1¢
r b w r b w r b w z
1 1 1 2 2 2 3 3 3 4

My generalization

I think that no matter how many of the same color you need you can always find out what the maxime cost will be You just figure out how many gumballs with the same color is needed. For example if the amount you need is ten the same h

You need to figure out how many different colors are inside the machine. You take the number you need (in this case ten) and multiply it by the number below the number you need (it would be nine) The answer is 90 Then it doesn't matter what color comes out it will match to make a ten. All it takes is one more making it 91

Example of work:

number needed –
one less –

10
x9
——
90

⟶ multiply

any color from machine –

90
+1
——
91

⟶ add one to the 90

continued>>

Student F

(4)

3 colors of gum } # of kids   1   2   3   4   5   n   otherterm
¢   1¢   4¢   7¢   10¢   13¢   -2+3n   -2

2 colors of gum } # of kids   1   2   3   4   5   n   otherterm
¢   -1¢   3¢   5¢   8¢   11¢   -1+2n   -1

formula → x = # of kids
y = # of different colors

$$(1-y) + yx$$

<u>Process</u>: I solved the problem by making two in-out tables. These tables gave me different chunks of information to put together. The first was used to find the relationship between the number of kids money. The second was used to find the relationships between the kids, money, and number of different colors of gum balls. The hardest part was starting because I didn't know exactly.

For parents with more than two children, we will use a visual representation. In order the require a maximum number of gumballs, the machine must first produce one gumball of each color, then a second of each color, a third of each color, and so on. If there are 4 colours in the machine, and 3 children, the max # of gum balls can be represented as:

color   1   2   3   4

        1 2 3 4

        5 6 7 8

        9

each row gets completely filled before a new row is started. As soon as there is one gum ball in the final row, the third row for the third child, there will be the same color of gumball with 3 of that color, one for each child. Each row before the row for the final child must be completely filled, meaning (the number of ~~colors must be bought,~~ children − 1) times the number of colors must be bought, +1 final gumball for the last child. The formula is ∴ max \$ = $(K-1)g + 1$

     where   K = # of Kids

            g = # of colors

continued>>

In order to generalize, group the gumballs in sets of colors so we have a machine with three colors, red, green, and blue, in the worst case scenario you will draw R, G, B, or some different order of these colors. If you have 3 kids, you will keep drawing and again receive a set of R, G, B. Then the next one you draw, no matter what color, will catch up with one set of two colors you previously drew, giving you three gumballs of the same color. So, you have drawn out n-1 sets of gumball colors, and you draw 1 more giving you $(n-1)g+1$.

Debriefing Activity 6.3: Analyzing Student Work From the Sticky Gum Problem

The Sticky Gum Student Work Analysis chart (shown in Figure 6.5) provides our analysis of each of the eight arguments through the lens of the Reasoning-and-Proving Framework. The column headings of this matrix are a little bit different from your analysis chart, so the information in Figure 6.5 extends the analysis you did in Activity 6.3. Most notably, we added a column in which we summarized the approach taken by the student, and we collapsed proof and non-proof arguments into one column. After reading through the chart, take a few moments and compare our analysis with your analysis.

FIGURE 6.5 Sticky Gum Problem student work analysis chart.

	Summary of Student's Approach	Identifies a Pattern	Makes a Conjecture	Provides a Non-Proof Argument or a Proof
Student A	Uses tables to explore numeric relationships between the number of children, colors, and cost. Builds three tables keeping the number of children constant (2, 3, 4) while varying the number of colors to find cost.	Total cost can be found by multiplying the number of children by the number of colors and subtracting the previous number of colors. Recursive pattern.	Creates an equation that "works" for the 15 examples in the table. $n = xy -$ (previous y) Variables undefined.	There is no justification for why the generalization works for all cases. *(Empirical Argument)* The development of the proposed equation is based on the examples in the table and not the problem context.
Student B	Uses tables to explore numeric relationships without evidence of considering the context. Builds two tables keeping the number of children constant (2, 3) while varying the number of colors to find cost.	Identifies patterns within each of the tables but does not construct a generalization.	Creates expressions for finding the cost when there are two children $(x + 1)$ and three children $(2x + 1)$, where x is the number of different candy colors.	There is no generalization. No evidence of a connection between problem context and table. *(No Argument; makes progress toward a generalization)*

(Continued)

	Summary of Student's Approach	Identifies a Pattern	Makes a Conjecture	Provides a Non-Proof Argument or a Proof
Student C	Creates an argument based on the context of the problem, but focuses on the specific example of three children and three gumball colors.	Solution is focused on one example.	Creates a generalization: $xy - (x - 1) = z$, where x = colors, y = children, and z = cents. More is needed to "define" variables.	This argument could be mistaken as being general and/or a generic example proof, but the discussion of the generalization does not go beyond the specific example. *(Empirical Argument)*
Student D	The student seems to have used guess and check until he/she identified the equation. There is a focus on generalizing from the table of numbers.	Identifies a pattern between the number of children, colors, and cost in the table, but does not relate pattern to context.	Creates a generalization. $(K - 1)C + 1 = A$ Does not define variables.	There is no argument provided.
Student E	The student writes that you need to find out how many gumball colors there are in the machine, but does not use this value, instead claiming that you multiply by one less than the number "needed" or number of children.	Identifies a pattern between the number of gumballs you need and the number 1 less than the number of gumballs needed.	Does not use symbols to express the identified relationship.	Provides some explanation regarding why the identified relationship works. It is unclear if the student is only considering the case where the number of colors and number of children are the same, or if the student is ignoring the number of gumball colors and multiplying the number of children by one less the number of children. *(Unclear Argument)*

	Summary of Student's Approach	Identifies a Pattern	Makes a Conjecture	Provides a Non-Proof Argument or a Proof
Student F	The student creates tables to explore numeric relationships between the number of children, colors, and cost. Each table keeps the number of colors constant (2 and 3) while varying number of children to find cost.	Identifies patterns within each of the tables but does not provide an argument for the generalization.	Creates a general expression for any number of children and any number of colors, $(1 - y) + yx$, where $x =$ the number of children and $y =$ the number of different colors.	There is no argument for the generalization. This student writes words just to say what is in the table as opposed to justifying why the generalization works or how he or she reached the generalization. You may notice the difference between what Student C and Student F wrote.
Student G	Constructs an argument based on the context of the problem that includes a model of the situation. Includes general language that goes beyond the specific case.	Does not show work indicating that a pattern was explored.	Creates a generalization for any number of children and any number of colors: $(k - 1)g + 1 = $ max, where $k =$ number of kids and $g =$ number of colors	Provides a coherent explanation that connects to the context of the problem. Provides a *generic example proof*—general language is included.
Student H	Creates a logical argument based on the context of the problem.	Uses a specific case to explain the general case.	Creates a general expression for any number of children and any number of colors: $(n - 1)g + 1$, where $n =$ number of kids and $g =$ number of colors	Provides a coherent explanation that connects to the context, however, the argument is based on a single case. Unlike Student G, there is no general language justifying the generalization. This is similar to Student C. *(Empirical Argument)*

One interesting thing to note about this set of student work is that only Student G provided what we consider to be a proof of the generalization. This student begins with a specific example and then discusses the mathematical structure of that example. Student G then argues that this structure will occur for all of the cases, thus providing a generic argument, one of the two types of proof we presented in Chapter 3.

We also want you to note that even though the other seven students did not present a proof, they all engaged in aspects of reasoning-and-proving. For example, Student C provides a solid argument for one particular case, but does not argue the generic case from this empirical argument. Student H does something similar. If the teacher engaged all of these students in a whole-class discussion of their solutions, these two students might realize the limitations of their arguments and be able to add text about the general case. And, as we mentioned in Chapter 4, Knuth (2002a) reminds us that "classroom presentations of different arguments can furnish a forum for discussing with students the question of what constitutes proof—a notion for which many students have an inadequate understanding" (p. 489).

You may have also noted how many of the students relied on multiple representations to make sense of the situation and explain their thinking. Another interesting thing is that the students who were depending on tables of values to make their generalizations (A, B, D, and F) had difficulties constructing a proof argument—they seemed to have lost a connection to the problem's context. The evidence of this is based on what the students wrote as an explanation. For example, Student F simply wrote some thoughts about the two tables, such as, "I solved the problem [by] making in-out tables." On the other hand, two students (C and G) who drew diagrams were able to connect the context to their generalization. In other words, the diagrams supported the students with envisioning the problem context and providing a clearer explanation of why they believed their generalization held true for all cases.

ANALYZING TWO DIFFERENT CLASSROOM ENACTMENTS OF THE STICKY GUM PROBLEM

In Activity 6.3, you will analyze two different enactments of the Sticky Gum problem. Calvin Jenson and Natalie Boyer each implemented the Sticky Gum problem in their respective algebra classrooms. Calvin and Natalie are known in their schools for wanting to create a learning environment for their students that supports sense-making, and they each selected this problem because they felt that the context of the Sticky Gum problem would provide another resource to support students' sense-making. From the outset, both teachers had good intentions for implementing the Sticky Gum problem with their students.

Activity 6.3 Comparing Two Enactments of the Sticky Gum Problem

1. Read *Making Sure That All Students Understand: The Case of Calvin Jenson* (Appendix E) and *Helping Students Connect Pictorial and Symbolic Representations: The Case of Natalie Boyer* (Appendix F) and identify instances where Calvin and Natalie supported or inhibited their students' learning.

2. In your journal, describe the nature of Calvin and Natalie's use of each of the five focal teaching practices:

   a. Establish mathematical goals to focus learning

   b. Implement tasks that promote reasoning and problem solving

   c. Use and connect mathematical representations

   d. Facilitate meaningful mathematical discourse

   e. Pose purposeful questions

Debriefing Activity 6.3: Comparing Two Enactments of the Sticky Gum Problem

Calvin Jensen and Natalie Boyer both wanted their students to have the opportunity to engage in reasoning-and-proving and selected a task that had the potential to engage them in these practices. They each supported students by allowing them to work in groups, helping students understand the meaning of "worst-case" scenario, and interacting with students while they worked on the task. In Figure 6.6, we provide additional actions the teachers took to support students' learning.

Despite the support they provided, both teachers implemented the task in ways that took away students' opportunities to make sense of the situation. Calvin began the lesson by making a chart of what was given in each part of the problem rather than letting students

first grapple with the task. Natalie revised the task and, in so doing, limited the students' opportunities to examine a wider set of cases. Hence, each teacher began the class by limiting students' opportunities to grapple with the original problem as stated.

In addition, Calvin spent too much time on the first question, and as a result, students had no time to actually engage in any of the practices related to reasoning-and-proving. Natalie pressed students to represent the generalization algebraically, but she did not push them to explain how the symbolic representation connected to the context of the problem. While they used specific examples to "test" that their formula worked, they did not make the argument that the generalization held true for all cases. As a result, Natalie's students may have been left with the notion that testing a generalization with different numbers is sufficient for proving that a generalization is true for all cases. In Figure 6.6, we summarize ways in which Calvin and Natalie supported and inhibited their students' learning and include additional actions you might have noted in your analysis.

FIGURE 6.6 Actions taken by Calvin and Natalie that supported or inhibited students' learning.

	Support	Inhibit
Calvin Jensen	• Provided an opportunity for students to work in groups (lines 33–34). • Provided students with cubes to represent gumballs (lines 30–31). • Promoted student talk (lines 35–38, 49–50, 77–78). • Suggested multiple representations (lines 45–47) • Asked the student-groups questions (lines 35, 49–50, 78). • Anticipated that students may misinterpret "worst-case scenario" (lines 28–30).	• Spent 15 minutes helping students understand the task and creating a chart rather than giving students the opportunity to make sense of the task themselves (lines 25–28). • Too much time—40 minutes—was spent answering Question 1 (line 112). • Did not provide the opportunity for students to individually think about the problem. Individual think time was replaced with the whole class creating a table of information (lines 25–29). • Seemed to focus exclusively on the use of cubes to explain the solution (lines 59–61, 78–80). • Too focused on making sure that every single student could explain the meaning of 2 and 3 cents, and as a result, little progress was made (lines 38–45, 52–70, 86–102). • Started his interactions with the second group with a question instead of trying to learn what they understood (lines 49–51).

	Support	Inhibit
Natalie Boyer	• Provided opportunity for students to work in groups (lines 28–29) • Monitored a few groups' thinking before jumping in and asking questions (lines 29–34) • Noticed that multiple groups had trouble with worst-case scenario and addressed this as a whole-class issue instead of discussing it at each table (lines 28–35) • Made sense of how students represented information (two different representations) (lines 55–60, 90–100) • Worked toward a generic example, connecting generalization to a specific example (lines 73–77, 100–104)	• Revised task to limit the opportunities to examine cases (removed three children and three colors) (lines 17–19). • Led students to symbolically represent the situation. Lee may have been confused about how his work transformed to $(k - 1)c + 1$. Words may have been an easier transition (lines 65–76). • Pressed for algebraic representation but not for explaining why this always works—not clear what Lee's group understood. $c + c + c + 1$ to $3c + 1$ to $(k - 1)c + 1$ may be too big of a leap (lines 71–77). • Focused on checking to see if the expression holds true instead of asking the students to explain why the expression always works or doesn't. Push for inductive instead of deductive reasoning. Pick a number of colors and number of children to check the expression (lines 103–110). • The teacher explained how Lee's formula connected to a specific example. Two issues: A specific example is not generic, and the teacher did the explaining (lines 71–77).

In the next sections, we debrief the cases through the lens of the five focus teaching practices from *Principles to Actions: Ensuring Mathematical Success for All Students* (NCTM, 2014). As you read our analysis, you may feel that we are being overly harsh with our comments. We do not intend this analysis as a criticism of Calvin and Natalie as teachers. Both of these teachers were committed to providing reasoning-and-proving and sense-making opportunities for their students, and the lessons portrayed in the cases represented their initial efforts to provide those opportunities for their students. What these cases make salient, however, is that sometimes well-meaning teachers do things that limit students' opportunities to fully engage in the range of reasoning-and-proving activities. Our purpose in asking you to analyze these two cases of "sometimes well-intentioned teaching goes awry" is to encourage you to look closely at how teachers interact with students in both productive and not-so-productive ways, as well as to consider the impact those interactions can have on student learning. Be sure to keep our learning goal for you in mind as you read the following analyses.

Establish mathematical goals to focus learning. As we discussed in Chapter 5, learning goals should describe what students are to learn about mathematics as a result of engaging in a particular lesson. They provide guidance for teachers as they ask questions, make decisions about which solutions to present, and help students make connections between different solution strategies and to the key ideas being targeted during the lesson. Without clear and explicit learning goals, a teacher has no clear target for instruction and can end up focusing on task completion without specific attention to what was being learned in the process.

In the opening of the case (lines 9–14), Calvin Jensen indicates that his goal for the lesson was "for students to determine the maximum amount that Mrs. Hernandez would have to spend to satisfy the children in each of three situations: twins and two gumball colors, twins and three gumball colors, and triplets and three gumball colors." His goal focused on what students would do rather than what they would learn related to reasoning-and-proving, suggesting that this goal is a performance goal rather than a learning goal. However, by focusing the entire class only on the answer to the first question—why 3 cents is the most amount of money Ms. Hernandez would have to spend—he not only didn't accomplish what he set out to do, his students had limited opportunities to engage in reasoning-and-proving.

Natalie Boyer does not explicitly state her goals for the lesson. In reading the case, it would seem that Natalie wants her students to create a generalization that represents the context. Her work during the class was clearly focused on helping students develop a generalization. Natalie seems to focus on what students will do (performance goal), but it is less clear what she wants her students to understand and learn about reasoning-and-proving (learning goal).

In both cases, the lack of clear goals for student learning may have contributed to the outcome of the lessons. Without clarity regarding what students were to learn related to reasoning-and-proving, both lessons ended up focusing on getting answers rather than on engaging in looking for patterns, making conjectures, and creating arguments.

Implement tasks that promote reasoning and problem solving. While the task provided ample opportunities to engage in a wide range of activities related to reasoning-and-proving, the way that Calvin launched the task may have been problematic for some students and thus limited their opportunities to engage in reasoning-and-proving. For example, introducing the term "worst-case scenario" at the beginning may have been confusing to students. Calvin's students understood that 2 cents and 3 cents were possible solutions to the problem, but perhaps they did not think that spending 3 cents is such a horrible situation. Perhaps considering a worst-case scenario was a stumbling block for students, or, perhaps, Calvin spent far too much time drilling students on this point. It is clear, however, that Calvin's launch of a high-level task could account for his students' lack of progress. Alternatively, during the launch of the task, he could have suggested that students focus on the most or least you would spend in any situation. This suggestion may have allowed the students to make more progress toward developing a mathematical argument.

We also think that Natalie's implementation of the task was problematic in supporting her students to engage in reasoning-and-proving. In the whole discussion at the end of the case, Natalie was trying to connect the students' generalizations to the context through the use of the organized list she had written on the board, which is a positive pedagogical move. However, we think that in this class, Natalie did too much of the mathematical thinking in making that connection. Because she did not require her students to do the mathematical thinking, it is not clear what sense students were making of the generalizations and how they connected to the context. As a result, it appears that students learned little about reasoning-and-proving from engaging in this task.

Additionally, Natalie's revision of the problem—eliminating exploring triplets—may have made it more difficult for students to engage in reasoning-and-proving. When students are examining cases to find patterns that could lead to making a generalization, they require *more* opportunities to explore examples, not fewer. Also, with the three variables, students may have benefited from keeping one variable constant to explore change in the other variable, which was not evident in their exploration. Natalie promoted a generalization from a single case as opposed to a generalization from a pattern that emerged from considering multiple cases.

Teaching Takeaway:

Don't do the thinking for your students! If you are doing all of the mathematical thinking during class, you seriously limit your students' opportunities to learn.

Teaching Takeaway:

Often, students need to explore many examples in order to make sense of a context and create generalizations. Give your students time to explore!

Use and connect mathematical representations. Mathematical representations are intended to provide support to students in making sense of mathematics and developing arguments. An important caveat here is that the choice of representations to use has to make sense to the *learner*. Calvin provided cubes and encouraged students to use the cubes in their explanations. While physical representations like cubes can be helpful for some students in solving this problem, other students may choose to use a different representation. For example, Tamyra and Merek made a table of possible combinations.

We think that Calvin missed an opportunity to help students connect representations by not engaging with this group about their work. So, while the cubes may have been a critical support for some students, for other students, using them may have actually made the task harder. The pedagogical lesson learned here is that you want to make manipulatives available to students as a resource, but you shouldn't force them to use the representation that you pre-decided would work. Let students use representations that make sense to them. If they get stuck, you might suggest that they think about the situation through a different representation. You need to be prepared to assess and advance students' thinking *based on the representations the students use*.

Although Natalie did not provide physical manipulatives, students drew diagrams to support their thinking and Natalie made an effort to understand the meaning of their diagrams. However, Natalie did the cognitive work (the thinking) of connecting the diagram to the generalization instead of using advancing questions to move the students toward being able to write a generalization. For example, in lines 59–60, Natalie explains how she interpreted what Lee shared and how she wanted to "help" the group construct a generalization. Natalie is the one to write the "*c*," and she also wrote the addition signs. It is not clear how the other students in the group were thinking about Lee's diagram, or what their thoughts were with respect to constructing a generalization. Lynn suggested an expression, and Natalie asked Andy and Lee to "test it." It is unclear what Andy or Lee understand about the generalization and testing it would not support them in understanding how the two representations were connected or in constructing their own argument.

Pose purposeful questions. Calvin's questions did not seem tailored to the progress students were making on the task, which conflicts with the idea of posing "purposeful" questions. For example, he asked a similar question when he engaged with the first two groups and continued to focus on Question 1 of the task with the third group even though the group had moved beyond the first question. Overall, he tended to ask questions without first listening to what the students understood; how they were making sense of the problem situation; and whether they were, in fact, struggling with his anticipated difficulty. He did not use assessing questions effectively.

Natalie, on the other hand, observed what students produced and asked students to explain. But at times, Natalie did too much of the "heavy lifting." For example, Natalie explained to her class the connection between the generalization and the student-generated example (lines 65–76, 106–110) rather than asking students to provide the explanation—she told the students what the connection was as opposed to allowing them to figure it out and explain it in their words. This was also evident when Natalie posed the question about how Christopher and Lynn's expressions worked based on the picture. The problem seemed to be that the students did not understand how the diagrams connected to the generalization. Therefore, when the question was met with silence, Natalie took it to mean that she needed to show them the connection between the diagram and the generalization. So, while Natalie effectively used assessing questions, she could improve how she uses advancing questions to support her students to move toward the goal of writing a mathematical generalization.

Facilitate meaningful mathematical discourse. Facilitating discourse does not occur only during whole-group discussion. In Calvin's case, he facilitated discourse in the small groups. Prior to teaching the lesson, Calvin anticipated what he thought might be difficult for students and prepared questions to ask them. Each time that Calvin engaged with a small group, he seemed to follow one "script" where he started or promoted a demonstration using the cubes. Then, after he left a group, it is unclear if the students made any additional progress on the problem or if they were more confused. He already had the answer to Question 1 on the board and still discussed this question with all three groups. While Calvin was consistent in engaging all the members of the group, he did not seem prepared to support students based on their current thinking; instead, he was only prepared to "listen" for a specific response, which seemed to slow student progress. This is problematic toward making progress with facilitating meaningful mathematical discourse since the teacher seems to be funneling students to explain the situation in a particular way. Instead, *Principles to Actions: Ensuring Mathematical Success for All* (NCTM, 2014) recommends that when you engage with students in a group as they work on a high-level task, you want to assess students' current understanding and learn how the individuals in the group are making sense of the problem from their perspective. Additionally, you want to also make note of student methods as you monitor their small group work in order to support the orchestration of your class discussion similar to what Nancy Edwards modeled in Chapter 4.

While Natalie was walking around helping groups, she selected three students to list their expressions on the board to set up the class discussion. When the discussion began, however, she introduced a new generalization to work from rather than using what the students had already produced. She used choral response to create a diagram that was similar to what Lee produced and walked the students through the connection between the symbolic representation (the generalization) and the pictorial representation (the diagram). After creating the diagram, Christopher volunteered another generalization ($ck - c + 1$), which was a version of what Lee's group had discussed and that Lynn had recorded. Natalie verbally connected Christopher's expression with the one that Lynn had written on the board. When Natalie asked the students to explain the connection between Christopher's and Lynn's different-looking expressions, she was "met with silence." So, Natalie went on to explain the connection between the two expressions. Also, there was never any discussion regarding the limitations of the first two expressions ($k + c - 1$ and $kc - 1$). Although Natalie's students ended up with two correct generalizations, it is not clear that students in class understood how they connected to the problem. Based on where these students were mathematically, more discussion was needed to better support them with learning how to construct a generalization and how to prove it.

In both cases, the opportunity to engage in reasoning-and-proving was not fully realized. From the analysis of these two specific cases (and the others you encountered in earlier chapters), you can continue to deepen your understanding about pedagogical practices that support student learning opportunities.

Return to the page in your journal that you started in Chapter 2 and added to in Chapter 4, titled "Pedagogical Moves That Support Students' Capacities to Reason-and-Prove." Reflect on the positive pedagogical moves that Calvin and Natalie made during these lessons and add them to your list.

Find (or modify) a task that provides students with the opportunity to engage in some or all aspects of reasoning-and-proving and is set in a context such as the Sticky Gum problem or the Building a Staircase task.

Part 1: Plan the Task

As you plan, reflect on these questions and prompts:

CONNECTING to Your CLASSROOM

1. The Sticky Gum problem and the Building a Staircase task are both "low floor high ceiling" tasks. To what extent does the task you identified provide access for all students and build on prior knowledge? To what extent does your task challenge students? What changes (if any) do you need to make to the task that enhance students' opportunities to reason-and-prove and don't lower the cognitive demand of the task? What additional opportunities are afforded by the modifications?

2. What *representations* will support students' work on the task? What representations might *you* need to introduce to students either prior to or during their work on this task?

3. Anticipate how you think students will solve the task. If you cannot generate a number of different approaches, ask other teachers to solve the task. These solutions will help you anticipate the ways that your students are likely to approach and solve the

continued>>

<<continued

task. Looking across the various strategies, generate questions you can ask to assess and advance students' learning.

4. Review the variety of solutions that you anticipated and consider which ones are most likely to help you achieve your goals for the lesson. Determine a sequence of solutions that you want shared that would allow all students access to the ideas you are targeting and support them in making connections toward your identified mathematical goal. (Recall the way that Nancy Edwards selected and sequenced the students' solutions to the Sum of Two Odd Numbers problem.)

5. Plan how you will launch the task in a way that does not limit students' opportunities to reason-and-prove.

Part 2: Implement the Plan

Teach the lesson you have planned in one of your classes. Audio- or video-record the lesson if possible and collect student work and other artifacts that were produced during the lesson. Use the artifacts from your lesson, including your plan, to analyze your pedagogical practices and student learning.

Part 3: Analyze Your Pedagogical Practices and Student Learning

After you have taught the lesson, reflect on these questions. If possible, engage in a discussion with your colleagues.

1. To what extent did students engage in reasoning-and-proving? What evidence supports your claims?
2. Was the discussion you facilitated meaningful? Why or why not?
3. In what ways did your planning prior to the lesson impact the outcome?
4. How did your use of the effective teaching practices help support students' learning and engagement?

******

In this chapter, you considered the role of context in supporting students' engagement in reasoning-and-proving activities. Your work on the Sticky Gum problem provided a way to pull together many of the ideas related to the teaching and learning of reasoning-and-proving that have been introduced in this book. In particular, as you created a proof, examined student work, and analyzed episodes of teaching (all around a contextual task), you drew on the tools introduced in previous chapters to think more deeply and critically about reasoning-and-proving and the quality of arguments and teaching. Moreover, as you drew on these tools in completing the activities in this chapter, you considered the role that context plays in supporting reasoning-and-proving. We hope that you have seen how context can be more than a source of data to be abstracted and then abandoned—that it can actually help in making sense of mathematics. As we move into the final chapter of the book, we suggest that you take a few minutes to reflect on the ways that you have deepened your knowledge about and capacities to engage in reasoning-and-proving as a result of the work you have done in this book.

Discussion Questions

1. Describe how you used context to support your work on the Sticky Gum problem.

2. In Activity 6.2, Student D created a correct generalization for the relationship between children, colors of gumball, and cost. However, this did not "count" as a proof. How could you move Student D toward writing a proof based in the way that he or she is already thinking?

3. Both Natalie Boyer and Calvin Jensen used a high-level task that provided the opportunity to engage students in reasoning-and-proving. Yet the task did not reach its potential in either class. Select one of the two teachers and articulate specific pedagogical moves that the teacher could have done differently that may have led to a different outcome.

Pulling It All Together

You brought to the reading of this book conceptions of what proof is, why it is or is not useful, and its role in the secondary school mathematics curriculum. These conceptions resulted from personal experience as a learner and teacher of mathematics and may not have always been positive. For example, you may have thought that proof had a limited utility in secondary school mathematics, that many students cannot do proof because it is too hard, or that there is no time in the curriculum to do proof given all the testable material that needs to be covered. We hope that as a result of reading this book, engaging in the activities, pausing and considering ideas that were raised, and making connections to your own practice, you now think differently about reasoning-and-proving. In this chapter, you will be asked to consider

- your current view of reasoning-and-proving; and
- how it is different from your initial view of proof.

KEY IDEAS AT THE HEART OF THIS BOOK

We have presented quite a few ideas about reasoning-and-proving throughout the book. In this section, we summarize key ideas about reasoning-and-proving that we have tried to communicate throughout the book. As you read through the list, consider how the ideas we have identified are reflected in your current view of reasoning-and-proving.

- **All students can engage in reasoning-and-proving given the appropriate opportunity and support.** For example, A. Stylianides (2007) provides an example of third-grade students providing a proof to the conjecture that the sum of two odd numbers is always even. While the proofs produced by third graders may differ in the use of representations (e.g., none used algebra), they adhered to the criteria for proof and they served to help students make sense of mathematical relationships. By providing all students with the opportunity to engage with the full range of reasoning-and-proving activities (i.e., identifying patterns, making conjectures, and providing arguments that may or may not qualify as proofs), more students will have access to new ways of reasoning and sense-making.

- **Reasoning-and-proving are mathematical processes that transcend mathematics content.** While reasoning-and-proving serves a valuable function in geometry classrooms, students should have the opportunity to experience it more broadly. There are ample opportunities to engage in reasoning-and-proving activities when studying geometry as well as algebra, statistics, and probability. The *Focus in High School Mathematics* series (http://www.nctm.org/store/fhsm/) and the Reasoning and Sense Making Task Library (http://www.nctm.org/Standards-and-Positions/Focus-in-High-School-Mathematics/Reasoning-and-Sense-Making-Task-Library/) provide resources for reasoning-and-proving activities in these content areas.

- **Proofs can take different forms,** which include, but are not limited to, two columns with reasons and supporting statements. Flowchart proofs (see Figures A6 and A7); proofs by exhaustion (Solution I in Figure 3.1); generic arguments, which begin with an example and argue that the mathematical structure of the example holds for all cases (Solution A in Figure 3.1); and narrative arguments (Solution F in Figure 3.1) are also appropriate ways to prove that something is always true. The key to a proof is not the form that it takes but rather that it shows the truth of a statement for all cases; draws on agreed-upon definitions, postulates, properties, and theorems; uses mathematics correctly; and has a logical flow (see Figure 3.3).

- **Using and connecting mathematical representations is an important component of reasoning-and-proving.** Pictures or diagrams can be a particularly compelling way to *represent* a proof because the visual images can help the viewer gain insight into why the conjecture is correct by "seeing" how the idea is illustrated geometrically or diagrammatically (e.g., Table 4.1, Solutions 1, 2, and 3; Figures B3 and B5). According to Nelson (1997, back cover), "Pictures or diagrams can help the reader see *why* a particular mathematical statement is true and also to see *how* to begin to go about proving it true." Hence, visual diagrams are a useful tool for representing and communicating proofs. Contexts are also important—not only do they aid in sense-making, but they also provide support for writing proofs.

- **Successful lessons do not just happen serendipitously—they are the result of thorough and thoughtful planning prior to the lesson, and also of implementing effective teaching practices during the lesson.** While you may have concluded that Gina Burrows, Nancy Edwards, and Vicky Mansfield were just good teachers who had motivated and attentive students, the reality is that these teachers were "good" because of the time and care they put into planning their lessons. They used the effective teaching practices to guide their planning efforts and engaged in these practices during instruction. In addition, as described in the Case of Nancy Edwards, they engaged in the practices of anticipating, monitoring, selecting, sequencing, and connecting to ensure that their discussions were, in fact, productive.

- **Doing the thinking for students does not support their learning or their abilities to reason-and-prove.** Providing students with a strategy to use or a rule to follow in solving a task may help them get an answer, but it has no residual impact. If students have not had the opportunity to make sense of the rule or strategy they are using, it is doubtful that they will retain it over time. This was made clear in the cases of Charlie Saunders and Natalie Boyer. Students who learn a rule without any conceptual base are likely to forget it completely, remember it incorrectly, or have no idea when to use it. Identifying patterns, making conjectures, and providing proof arguments can be an effective method for understanding why mathematical rules work the ways that they do.

TOOLS TO SUPPORT THE TEACHING OF REASONING-AND-PROVING

It is our hope that the expanded view of reasoning-and-proving we have described in this book will help you in building students' capacity to engage in these processes. The tools we have presented should make integrating reasoning-and-proving into your curriculum a less daunting task by providing direction and guidance as you continue to make these processes an ongoing feature of what it means to learn and do mathematics in your classroom. These tools are as follows:

- The **Reasoning-and-Proving Framework** (Chapter 3) defines the major activities involved in reasoning-and-proving (mathematical component) and draws attention to student perceptions (learner component) and teacher actions (pedagogical component). It can serve as a guide for characterizing activities in terms of their reasoning-and-proving potential and a reminder to consider students' perceptions and develop strategies based on those perceptions for moving students forward in their understandings.
- The **criteria for judging whether an argument counts as a proof** (Chapter 3) is a set of criteria that can be used by a teacher or students in judging when a mathematical argument counts as a proof and for determining what would be needed in order to enhance a non-proof argument so that it becomes a proof.
- The **effective mathematics teaching practices** (Chapters 4 and 5) describe purposeful actions teachers can take to support students' engagement in and learning from reasoning-and-proving activities. These practices—establish mathematics goals to focus learning, implement tasks that promote reasoning and problem solving, use and connect mathematical representations, facilitate meaningful mathematical discourse, and pose purposeful questions—are central to developing students' conceptual understandings more generally and their ability to reason-and-prove specifically.
- The **strategies for modifying existing tasks** (Chapter 5) are a core set of strategies that can be used to adapt existing tasks in ways that increase their potential to engage students in reasoning-and-proving. These strategies make it possible to use any curricular resource as a source for activities in which to engage your students.

PUTTING THE TOOLS TO WORK

In Activity 7.1, we ask you to return to the teachers at Hoover High School, whom we first visited in Chapter 1, and provide advice that will help move them forward in their quest to incorporate effective learning opportunities for their students to get better at reasoning-and-proving.

Debriefing Activity 7.1

As we have noted previously, the algebra teachers at Hoover High School were committed to improving their students' abilities to think deeply about mathematical concepts and relationships, reason mathematically, and write valid mathematical arguments. We commend their commitment to these goals. These teachers came together in a PLC to work on their teaching and developed a common goal for improvement that was based on analyzing student data. They chose the same task to implement in their classrooms and gathered artifacts about interesting things that happened during class. They each came away from their class with questions to discuss with each other. Despite their best intentions, these teachers struggled to support their students when they did not immediately follow the expected pathway.

In the bulleted paragraphs that follow, we describe what we see as the reasoning-and-proving dilemma each teacher faced and make suggestions on what ideas from this book might help his or her students. Following these bulleted paragraphs, we present a possible plan for moving the teachers' PLC forward.

- Carly Epson's student Shonda's abilities to write a proof were limited because she had not been confronted with the limitations of making mathematical arguments based on using only empirical examples. She would benefit from engaging in the three-task sequence (Chapter 2). Once that had been established in Carly's classroom, teaching her students about generic arguments (Chapter 3) would give Shonda a method for writing a proof that is appropriate at this stage in her mathematics education.

- Jason Steiner's student Keisha had not engaged in seeking patterns and would benefit from working with empirical examples and identifying patterns from which to build a conjecture. Perhaps modifying the task (Chapter 5) to make it more exploratory (as opposed to stating the conjecture) would have supported this kind of work for his students. Jason's students would also benefit from knowing about generic arguments (Chapter 3).

- Barbara Law's students Michael and Marissa were right on target—they had created a demonstration proof (Chapter 3) based on pictures. We can surmise that they had explored with a number of empirical examples and then were able to see the general relationship that described the pattern they noted. The dilemma here is that Barbara did not know if their argument counted as a proof, so her dilemma could have been solved if she educated herself more about the nature of reasoning-and-proving (by, for example, reading this book!).

- Although Lynn Baker was pleased with her students' abilities to prove the conjecture, we find it problematic that she did the initial thinking for them. We wonder what they could have done had she not set them up for writing the algebraic proof by defining the variables for them. The fact that every student did the same thing is suspicious to us, given the range of student work we presented in Chapter 2. Lynn, we believe, would also benefit from thinking hard about teaching in ways that support students to engage in reasoning-and-proving, as you did in Chapter 4.

MOVING THE HOOVER HIGH TEACHERS FORWARD IN THEIR PLC

There are a number of effective pathways based on the contents of this book that could be used to help the Hoover High teachers move forward in their PLC work. We present one such pathway here.

The teachers could begin by looking at all of the work produced by the students in all of their classes. They could collaboratively consider what the students currently understand and what they want them to ultimately understand, and then propose pedagogical moves to implement that might help move the students beyond where they currently are (the learner and pedagogical components of the reasoning-and-proving framework).

The teachers might then implement the three-task sequence (Chapter 2) so that students understand the need for proof and what counts as a proof. Following this activity, they could collaboratively construct a set of responses to the Sum of Three Consecutive Numbers task that included a range of different responses (like the set of responses to the Sum of Two Odd Numbers task in Figure 3.1), give the set of responses to students, and ask students which ones are most convincing. This could lead to having students generate the list of criteria for judging whether an argument counts as proof.

Finally, the teachers might select another reasoning-and-proving task to implement in their classrooms and plan the lesson together. The planning could begin by being clear what they want students to learn as a result of engaging in the task (establish mathematics goals to focus learning) and then selecting a task that would have the potential to accomplish the targeted goal (implement tasks that promote reasoning and problem solving). The task could require that students look for patterns and make a conjecture *before* engaging in creating a proof. This could provide more scaffolding for students and give the teachers a good sense about what students understand about the problem.

Once the teachers have settled on a goal and task, they could begin to anticipate how students would likely solve the task (both correctly and incorrectly), discuss the representations that students may use (use and connect mathematical representations), and generate assessing and advancing questions (pose purposeful questions) that they could ask students who produced specific solutions (see Table 4.2 for examples). Through this process, the teachers could work together to determine what will be challenging for students and how the teachers will help students move beyond the impasses that they encounter. The teachers could then consider what solutions could be shared with the whole class in order to accomplish their lesson goal(s) (facilitate meaningful mathematical discourse). The criteria for judging whether an argument counts as a proof could be used by students to determine whether or not the arguments they created did, in fact, rise to the level of a proof.

****

Changing your teaching practices is not easy. It takes time, effort, careful planning, and thoughtful reflection to improve teaching. Engaging in the activities in this book was the first step in a journey to improve your students' opportunities to get better at reasoning-and-proving. We hope that you will continue this journey beyond the reading of this book. We encourage you to identify colleagues (virtually, across town, or next door) and collaboratively support and encourage each other as you engage each and every one of your students in learning to reason-and-prove with confidence and understanding.

Discussion Questions

Over the course of reading this book, you encountered six different teachers, each of whom was committed to engaging his or her students in reasoning-and-proving activities. The teachers were

- Charlie Sanders (Chapter 2—the three-task sequence)
- Gina Burrows (Chapter 2—the three-task sequence)
- Vicky Mansfield (Chapter 4—writing flowchart proofs)
- Nancy Edwards (Chapter 4—the Sum of Two Odd Numbers task)
- Calvin Jensen (Chapter 6—the Sticky Gum problem)
- Natalie Boyer (Chapter 6—the Sticky Gum problem)

Reflect on the teaching and learning in each of these cases and consider these four questions:

1. With which teacher do you most identify? Why?

2. Which of the teachers do you most want to emulate? Why?

3. What will it take to get from where you are now to where you want to be?

4. How can the tools we presented in this book help you in making changes in your instruction?

Appendix A

Developing a Need for Proof: The Case of Charlie Sanders

Charlie Sanders teaches in an inner-city charter high school. With almost 30 years of teach-ing experience, he is the leader of the math department in his school. He and his colleagues want their students to understand the mathematics they are learning, and several years ago, they adopted a reform curriculum featuring challenging tasks that cannot be solved simply
5 *by applying a known procedure—students need to be able to think and reason about math-ematics. Charlie recently attended a regional conference of the National Council of Teachers of Mathematics (NCTM), where he heard several speakers talk about the importance of engaging high school students in proof across content areas and not just in geometry. This made him realize that he and his colleagues did not spend much time on proof outside of*
10 *geometry, and that they should try to provide more opportunities for students to under-stand what proof really is and why it is important, and to get better at actually producing proofs. He shared these thoughts with his colleagues, and they seemed eager to work with him on enriching their students' experiences with proof. Since only the geometry textbook dealt with proof in any substantial way, the teachers agreed to browse different professional*
15 *journals for activities that would help them make progress on their goal.*

The teachers found several articles that included good activities they could use, but they ended up choosing to focus, as a first step, on a sequence of activities discussed in one of the articles (A. Stylianides, 2009). They chose this particular sequence of activities for several
20 *reasons. The activities*

- *were outside the domain of geometry, and there was a detailed discussion of each activity in the article;*
- *were accessible to different groups of students, so all the teachers, regardless of grade*
25 *level, could try them in their classrooms with only minor modifications;*
- *aimed to help students see the limits of using only empirical arguments when proving conjectures; and*
- *formed an introductory sequence on proof, supporting future use of other proving activities that the teachers had found in their search.*

30

CHARLIE SANDERS TALKS ABOUT HIS PLANS

I decided to use the sequence of tasks—the Squares Problem, the Circle and Spots Problem, the Monstrous Counterexample Illustration, and the Squares Problem Revisited—with my 12th-grade calculus students. I have been with this group of students since they started as
35 freshmen through a process called "looping" (i.e., moving to the next grade level with your

students). Although this process is usually used in elementary schools rather than in high schools, it has worked well for our school in that it has allowed teachers to really know their students well both as people and as learners of mathematics. The students in this class have worked hard and they do well academically. But the truth of the matter is that we haven't done much with proof over the past 3 years, and they still have some misconceptions about when you have really proven something. My main goal for the activity sequence is to help my students start to realize that arguments based only on the examination of a few examples are not sufficient to prove that something is always true. I thought it would be good to use the sequence early in the year (it is the first week of October) so that it could provide a foundation on which to build over the next 8 months.

Although my students work productively in small groups on a regular basis, they work less well when sharing and discussing tasks as a whole class. I find that students are resistant to sharing their views publicly, fearing they may be wrong or criticized by other students. So, years ago I decided not to push students to present their solutions to the entire class or put students in situations where they may feel uncomfortable. Actually, the other math teachers in the school had similar concerns, and so we all decided that having whole-group discussions in which students "battle about mathematical ideas" is not something we would pursue. So, I do most of the talking in the whole-class discussions based on what I see happening in the small groups as the students work on an activity.

Below, I discuss how each activity in the sequence played out in my classroom. The implementation of the whole activity sequence was intended to take one 65-minute class period.

THE BEGINNING OF CLASS: THE SQUARES PROBLEM

I handed out a sheet that contained the Squares Problem (see Figure A.1) and asked my students to move to sit in groups and begin work on the problem. I found that by not assigning groups and allowing the students to choose their group members, the conversations among the students were more productive. That day, there were eight groups with the number of students in each

FIGURE A.1 The Squares Problem.

1. How many different 3 × 3 squares are there in the 4 × 4 square to the left?
2. How many different 3 × 3 squares are there in a 5 × 5 square?
3. How many different 3 × 3 squares are there in a 60 × 60 square? Are you *sure* that your answer is correct? Why?

 group ranging from one to four. Tim was working alone since Ben, his usual partner, was absent. In the past, I have asked students to find another group to work with if their partner was absent, but the students showed some resistance to joining other groups, so I no longer insist on it.

MONITORING SMALL GROUP WORK FOR THE SQUARES PROBLEM

70 As the groups settled in to begin work on the problem, I walked around the room to get an idea about what each group was doing. I expected that all students would construct a table and write an equation from the table in less than 5 minutes since they have had considerable experience with pattern tasks.

75 I stopped at a few tables to see what each group was doing. Just about all the students found that there were four 3 × 3 squares in a 4 × 4 square. Most groups were starting to make a table and labeling the variables correctly as shown below.

Square size	# of 3 × 3 squares
4 × 4	4
5 × 5	
6 × 6	

I moved on to Group 3, where I saw Gwen, Lilly, and John counting the number of 3 × 3 squares in a 5 × 5 square. Gwen started on the top left corner of the 5 × 5 square. Using her
80 finger to outline the 3 × 3 square, she counted one. Then she said, "Top left, bottom left, bottom right, top right, middle, top middle, bottom middle, left middle, and right middle. So, there are nine altogether" (as shown in Figure A.2—dotted lines represent what Gwen was showing on her paper). John and Lilly nodded in agreement as I moved across the room to Group 7. There, I saw Robert counting the number of 3 × 3 squares in the 5 × 5
85 square and observed him recording seven. Whitney, Robert's partner, seemed to follow his lead. Since this was not the correct answer, I suggested that they might want to check their count again. Then I moved on to Group 8. There, I witnessed two students (Shelia and Tom) drawing a 5 × 5 square using Microsoft Word on their laptops. The other two students in the group (Travon and Samuel) were debating as to whether there were eight or nine 3 × 3
90 squares in the 5 × 5 square. I thought to myself that this was taking longer than I had anticipated, and I wondered if the laptops were even needed for this task or if they were getting in the way of students making progress on the problem; it was taking too long for students to draw the squares using the computer.

95 I felt that I needed to speed things up a bit, so I walked to the front of the room to address the class. I explained, "I don't want you to get messed up counting squares," I announced.

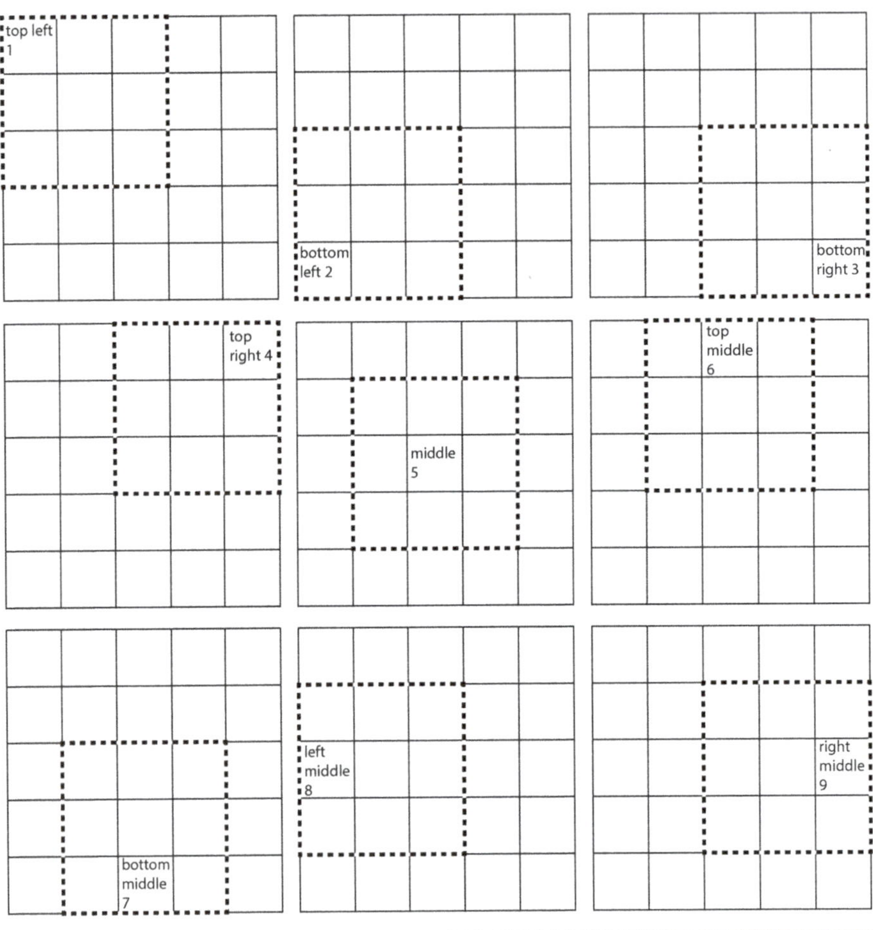

I asked, "How many squares are there in a 4 × 4 square?" I heard a student saying four, so I wrote 4 × 4 → 4 on the board. Then I asked, "How many in a 5 × 5?" I heard a few different numbers this time, including the correct answer, nine. In order to move the lesson along, I wrote 5 × 5 → 9 on the white board. With this scaffolding, I hoped that some students would begin to see a pattern, so I asked them to continue to work on the problem in their small groups.

Once again, I wanted to get a sense of the progress being made, so I checked in on several groups. I started with a visit to Group 7, where Robert and Whitney had 12 written for the number of 3 × 3 squares in a 6 × 6 square. I told them that they were not counting correctly and moved on. They have a tendency to be a little careless at times, so I wanted to make sure they were paying attention to what they were doing. Then I noticed that Tim, who was working alone, only counted 14 3 × 3 squares in a 6 × 6 square. However, Carlos and April (Group 1) thought they had a formula, and Shelia, Travon, Tom, and Samuel (Group 8) were also getting close to finding a formula. I listened in on Group 8 as they discussed the 60 × 60 square.

Shelia: You minus two from the number and double it.
Travon: No, you multiply it by itself.
Tom: So, I think we have it.
Samuel: Take the number, subtract 2, and then multiply it by itself.
Tom: Square it!
Travon: So the answer for Part 3 [in the Squares Problem] is 58 squared!
Samuel: [picking up the calculator from the table] That's 3,364 squares!

I was impressed that all the students in this group were working so well together and found the total number of 3 × 3 squares in a 60 × 60 square. Now that I had one group who had found the formula, I could discuss Part 3 of the problem with the whole class. I looked at the clock as I walked to the front of the room and noticed that we were just under halfway through our 65-minute class. I didn't expect the Squares Problem to take so long.

FACILITATING WHOLE-GROUP DISCUSSION ABOUT THE SQUARES PROBLEM

"A group found 3,364 3 × 3 squares for the 60 × 60 square," I announced to the class. Carlos shouted that he found a formula. "We made a table, then found an equation," he added. I looked at Carlos's work and copied the formula $(x - 2)^2$ on the board while asking if he had checked his formula. Carlos and April responded in unison, "Yes!" Looking to the back of the room at Group 8, I said that I thought a group in the back got something similar. "Samuel, did you get a formula?" I asked. "It was all of us," Samuel replied. "We got the same thing." Not wanting to spend much more time on this part, I asked how many students were certain that the formula would always work. "Who is willing to bet $100 that this formula will be true?" I asked. No student responded to my question, but I heard Mary whispering to a member of her group, "Why is he asking if we are sure about the formula? Why shouldn't we be sure?" I suspect that many other students would agree with Mary, especially given the certainty and the degree of excitement expressed by Groups 1 and 8 when they announced to the rest of the class that they had found a formula. Given that we were already running late, I decided to start the second task in the sequence, the Circle and Spots Problem. I told the class that some students came up with a formula and that the rest of them would gain a better understanding of it when we returned to the Squares Problem later in the class period.

INTRODUCING THE CIRCLE AND SPOTS PROBLEM

I distributed a statement of the Circle and Spots Problem (see Figure A.3) to students and explained a few basic points: (a) the lines could intersect, (b) "spots" meant points, and (c) "around the circle" meant on the circumference of the circle. I drew a circle on the

board and I began by recording the number of regions for one point, two points, and three
points with some student input about the number of regions formed when the points were
connected (as shown in Figure A.4). I suggested to the students that Geometer's Sketchpad
might be useful for their work on this problem. The students were comfortable using the
software since they use it at this school starting in ninth grade. A student asked, "Does it
matter where the points are?" I explained that it would be useful to space the points out
around the circumference of the circle so that the regions were bigger. All the students
opened their laptops and began drawing circles and connecting points.

FIGURE A.3 The Circle and Spots Problem.

Place different numbers of spots around a circle and join
each pair of spots by straight lines. Explore a possible relation
between the number of spots and the maximum number of
non-overlapping regions into which the circle can be divided
by this means.

*When there are 15 spots around the circle, is there an easy
way to tell for **sure** what is the maximum number of non-overlapping regions into which the circle can be divided?*

FIGURE A.4 Teacher set-up to the Circle and Spots Problem.

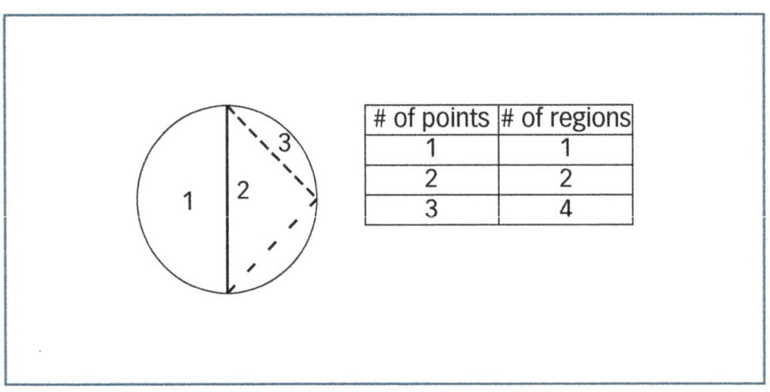

# of points	# of regions
1	1
2	2
3	4

MONITORING SMALL GROUP WORK FOR THE CIRCLE AND SPOTS PROBLEM

I noticed that several students started by placing 15 points on the circumference of the circle
and began using line segments to connect the points. As I walked by Group 8, I remarked,
"Good luck counting 15!" "Yeah, but it looks awesome," Tom replied. Tom's figure looked
very messy, and I doubted that it would be productive. I decided not to say anything more

   WE REASON & WE PROVE FOR ALL MATHEMATICS

to him because Samuel, a member of Tom's group, asked, "Will we ever be able to count all these regions?"

165 I then stopped at Group 6, where Tanya was starting to create a table like the one I started earlier (shown below).

# of dots	# of regions
1	1
2	2
3	4
4	8
5	14

I told her that 14 was not the number of regions for five points. I suggested that she talk to the members of her group and see what they had come up with.

170 I then returned to Group 3. Lilly's table was similar to Tanya's, but she had 16, the correct number of regions for five points. Gwen picked up her calculator and touched a few keys before writing the number 2048. She then turned to me and said that fifteen points would produce 2048 regions. I wasn't sure how she came up with this number, but I suggested that she check six points. After placing another point on her circle, she said that she only found
175 30 regions. The students in the group discussed how three lines intersected at the same point, and after adjusting a line, they found 31 regions. I then asked, "What does this do to your pattern in the table?" John remarked that according to the pattern it should be 32, but that they only found 31. As I understood it, the point of this task was for students to notice that the number of regions failed to double once the sixth spot was added to the circle. So,
180 when one of the small groups found 31 regions for six points, I felt I could bring the class together and talk about it. As I walked from the back of the room to the front to address the whole class, I quickly glanced down at a few more students' work. Group 8 had 31, so did Groups 2 and 5. I was happy to see that.

185 **FACILITATING WHOLE-GROUP DISCUSSION FOR THE CIRCLE AND SPOTS PROBLEM**

As I turned and faced the class, John said, "Patterns, are you trying to make the point that patterns don't always work?" I decided to ignore the comment because, although that was exactly the point I wanted to make, I felt it was a bit premature to make this point now. I
190 then showed a PowerPoint slide I had prepared to illustrate that six points would yield a maximum of 31 regions (shown in Figure A.5).

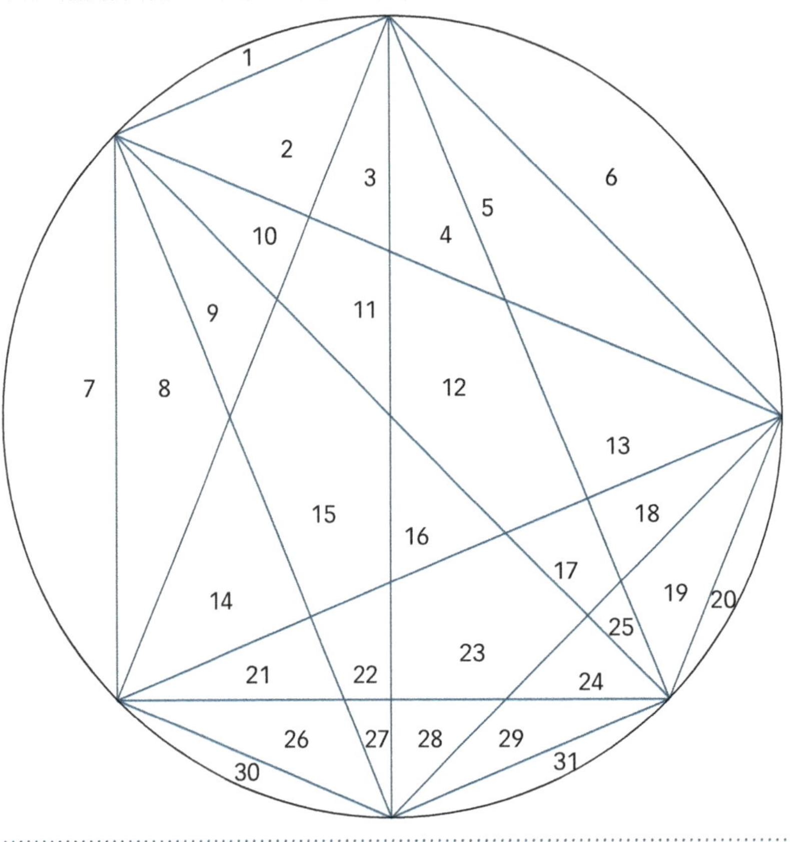

As the students looked at the handout with the regions numbered 1 to 31, I continued filling in the table on the white board I started previously (shown below).

# of points	# of regions
3	4
4	8
5	16
6	

After I wrote six points, I turned and asked the class, "Did anyone find a formula from this?" Heather indicated that she found 2 to the x minus 1. Not knowing how she wrote her equation, I wrote $2^x - 1$ on the board. Ted, who was a member of Heather's group, corrected me by saying, "No, $x - 1$ is the exponent." I rewrote the expression as 2^{x-1} and clarified with them that x represented the number of points. I then asked, "Does this work for the next one (six points)?" Students were in agreement that the formula did not hold for six points. Trying to motivate a need for proof back in the Squares Problem, I posed the questions: "Are you still sure that 3,364 is correct for the Squares Problem? Is anyone willing to bet $100 on that formula?" Some students conversed with members of their group, but none of the students responded to my questions publicly. As I mentioned earlier, my students are not willing to take risks in front of the whole class, so I didn't push them to respond.

INTRODUCING THE MONSTROUS COUNTEREXAMPLE ILLUSTRATION

I started work on the next activity by writing the expression $1 + 1141n^2$ and the constraint that n had to be a natural number on the white board. I turned to the students while pointing at the expression and said, "Someone claimed that this expression will never give a square number no matter which natural number n you plug in the expression. Can you find a number that will make the expression a square number?"

The students started working on this question in their small groups. Some students began to insert numbers to check, while others chose a different route. Carlos wrote the equation $1 + 1141n^2 = n^2$ and tried to solve for n. April explained to Carlos that he could not set the expression equal to n^2 because the number had to be different and suggested that he replace n^2 in the right side of the equation with y^2. Another group was trying to use the table function in their calculator. Samuel wrote the equation $1 + 1141n^2 = 4$ and attempted to solve for n. Another group began to substitute various numbers for n, such as 10, 36, and 852. I did not ask any questions, choosing instead to simply stroll up and down the middle aisle, making note of what the groups were doing.

FACILITATING WHOLE GROUP DISCUSSION OF THE MONSTROUS COUNTER EXAMPLE ILUSTRATION

Once I felt that the students had spent enough time searching for a solution, I gathered the class together for a discussion. I asked, "What is happening?" A student replied by saying that no number would make the expression a square number. I then took a poll of the class, and just under half of the 19 students agreed that no solution would work. Travon, who did not agree, commented that we do not know yet since we did not collect enough data. I then wrote the number 30,693,385,322,765,657,197,397,207 on the board and told students that this number would yield a square number. I commented, "Some of you were confident that your equation worked for the Squares Problem only after checking four or five cases. Then, after checking five cases in the Circle and Spots Problem, we found a formula. However, that formula failed on the sixth case. Travon pointed out that we needed to check more data, but this counterexample appeared in the thirty-septillionth case." I then concluded by stating, "Therefore, we cannot just check cases to prove in mathematics.

So now we will return to the Squares Problem to prove that the pattern we found there will *always* work." I may have rushed things a bit here, but I really wanted to at least start the last activity and assign homework before the end of the period.

240

SQUARES PROBLEM REVISITED

I then asked the students to return to the Squares Problem and to create a proof for the generalization. It was clear in a few minutes that students did not understand what I expected them to do. Nearly every hand was waving in the air trying to get my attention, and the

245   groups I checked in with all asked the same question: "What are we supposed to do?" I realized that they either did not understand what it meant to prove that a generalization was true or did not know how to go about proving this particular generalization. To help them, I distributed three different proofs that colleagues and I had created in order to illustrate how students could solve the problem.

250

I concluded the lesson by telling the students that, for homework, I wanted them to rewrite one of the proofs that I had given them in their own words. "Describe the proof in a way that you understand it," I explained. I figured that it would be a good idea to have the students explain one of the proofs rather than writing their own proofs since they were not sure

255   how to begin writing a proof.

As students quickly gathered their things and hurried from my classroom, I took a few minutes to reflect on how the lesson went in relation to what I read in the A. Stylianides (2009) article. I believe I closely followed the spirit of the recommendations in the article about the

260   implementation of the activity sequence. Also, although inevitably there were some differences since both the students and the teacher of the classrooms were different, the activity sequence played out in my classroom in a similar manner to how it played out in Kathy's classroom (the teacher in the article). I think my students got the point that you can't just rely on testing cases. The article does not say much about what happened in Kathy's class-

265   room after the class revisited the Squares Problem, so I need to think carefully about how I continue work on this problem tomorrow.

   WE REASON & WE PROVE FOR ALL MATHEMATICS

Appendix B

Motivating the Need for Proof: The Case of Gina Burrows

Gina Burrows has taught high school math for 19 years and is currently a mathematics coach at her school district's two middle schools. As a math coach, Gina regularly organizes after-school professional development sessions for the middle school teachers, so she is constantly researching new ideas to bring to the teachers to discuss. She recently read an
5 *article (A. Stylianides, 2009) about a three-task sequence that can be used to help students see the limitations of making valid mathematical claims based on finding a pattern from a few empirical examples. In her master's program, Gina has been reading about ways to help students become better at writing valid mathematical arguments. As she reflected back to her own teaching, she realized that she often allowed her students to depend on just a few*
10 *examples when making claims. She thought that middle school students would benefit from engaging with this sequence, and she developed a plan for making this a reality in her district.*

GINA BURROWS TALKS ABOUT HER PLANS

I decided that in order to be able to teach teachers about this three-task sequence and
15 convince them to implement it with their eighth-grade students, I needed to try it out with students first. I decided to enlist the help of Gavin Wyne, an eighth-grade math teacher at one of the district's middle schools. Together, we planned a 2-day lesson intended to help students recognize that empirical examples are not sufficient for making valid mathematical claims and arguments. I told Gavin that I would teach the lesson if he would video-record
20 it. We planned to share video clips and artifacts from the lesson with the teachers during an upcoming professional development session. I hoped that through this process I could encourage other teachers to try the task sequence with their students!

THE BEGINNING OF DAY 1—THE SQUARES PROBLEM

25 Because of my work as a math coach, Gavin's students were used to me being in the classroom and co-teaching. So, they were not surprised to see me there when they came into the room or to see Gavin behind the video camera. Gavin and his colleagues often film their lessons and share clips of videos during professional development sessions. As the bell rang and students settled in at their assigned tables and took out their notebooks, I called their
30 attention to the interactive whiteboard where I had projected the Squares Problem.

I started by telling them, "Today and tomorrow, we are going to work on a sequence of problems that will require you to find patterns and make conjectures about mathematical relationships. In the baskets on your tables, there are a number of items—graph paper,

35 scissors, markers, rulers, and calculators—that you can use as needed while you are working on the problems." I then passed out copies of the Squares Problem.

To launch the problem, I asked students to read it silently and then to turn to their elbow
partner and discuss what the problem would require them to do. Jamie raised his hand.
40 "Nadia and I don't understand what you mean by a 3 × 3 square. We see lots of squares
in the picture." In response, I went to the interactive whiteboard and drew a heavier outlined 3 × 3 square in the upper left corner of the 4 × 4 square. "Here is one example of
a 3 × 3 square in the 4 × 4 square. Who can come to the board and show us a different
3 × 3 square?" Donny raised his hand, and he came to the front and used a heavier line
45 to draw another 3 × 3 square in the bottom right of the 4 × 4 square (see Figure B.1).
"Jamie and Nadia, does that help?" They said yes. I asked if there were other questions
and paused. Seeing no hands, I continued.

FIGURE B.1 Drawings of 3 × 3 squares on the interactive whiteboard.

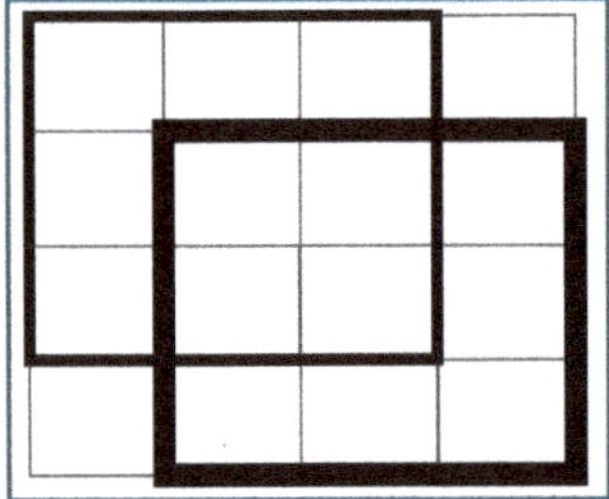

I then asked students to work individually for 3 minutes, and I set the timer on the interactive whiteboard. I think it is important for students to be able to think about a problem
50 individually before they start talking to their peers. It gives students a chance to start to
explore on their own, which I think brings more ideas to the table once they start sharing.
As those 3 minutes came to a close, I said, "OK, finish the thought that you are having and
then when you are ready at your tables, share what you've been thinking about with your
tablemates."

55

MONITORING SMALL GROUP WORK ABOUT THE SQUARES PROBLEM

As the students began to talk to each other, I paused at a table where Sandra was sharing what she had done so far with other members of her group. "Well," Sandra said, "I
decided to cut out a 3 × 3 square from the graph paper to use and shift it around. I spent a
60 lot of time doing that, and I was just getting to the point where I was going to start draw-

    WE REASON & WE PROVE FOR ALL MATHEMATICS

ing bigger squares and counting." Abby jumped right in and said, "I think that's a great idea. Can I do that too?" She looked up at me. I asked, "How do you think that might help you?" Abby pointed to her paper and said, "I've been drawing squares on the graph paper, but it's getting so messy that I keep losing count. I think having a cut-out square to

65   use would help." I turned to the third student, "Cody, what do you think?" Cody nodded, so I left this group to cut out 3 × 3 squares and proceeded to check in with other groups.

I noticed that Group 3 was staring intently at a paper in the middle of the table, so I stopped and said, "Tell me what you have been thinking about in this group." Justin said,

70   "Well, when we started talking, we all thought that Sean had a pretty good way of keeping track of the 3 × 3 squares when he was counting the big squares." I turned to Sean. "OK, explain what you did." Sean paused and then said, "Well, instead of thinking about the whole 3 × 3 square, I noticed that each 3 × 3 square has a middle square." I didn't quite understand what he meant, so I asked him to sketch the 3 × 3 square and show me (see

75   Figure B.2). I looked at it and then asked, "How is that going to help you?" I looked to another group member, "Pasma, can you explain?" Pasma pointed to another figure they had drawn (see Figure B.3) and said, "Look at this 5 × 5 square. The dashed square in the corner is our first 3 × 3 square, and the blue X is the center of that square." I nodded. "So," she continued, "if we just marked down the centers of the squares as we shift left

80   and then down and over, then we can just count up the Xs to find out how many 3 × 3 squares are in the 5 × 5 square." Not sure that I fully understood the process that Pasma was describing, I asked if another member of the group could explain the meaning of the nine Xs. Justin replied, "Those nine Xs are the centers of all of the 3 × 3 squares that are in the 5 × 5 square. Each X represents one square. We tried it another way and got nine,

85   too, so we think it is right."

FIGURE B.2 Sean's sketch showing the middle of the 3 × 3 square.

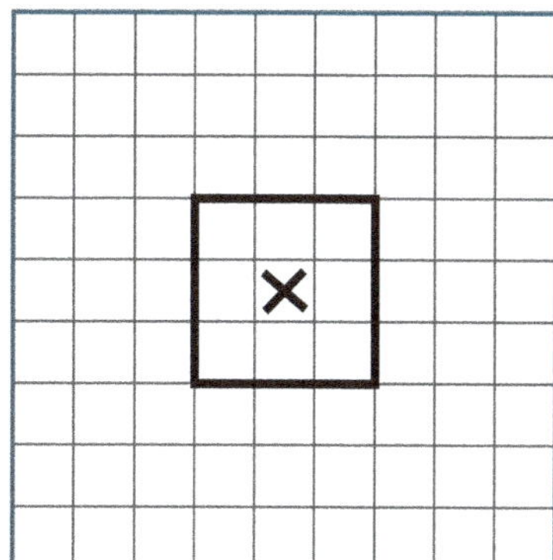

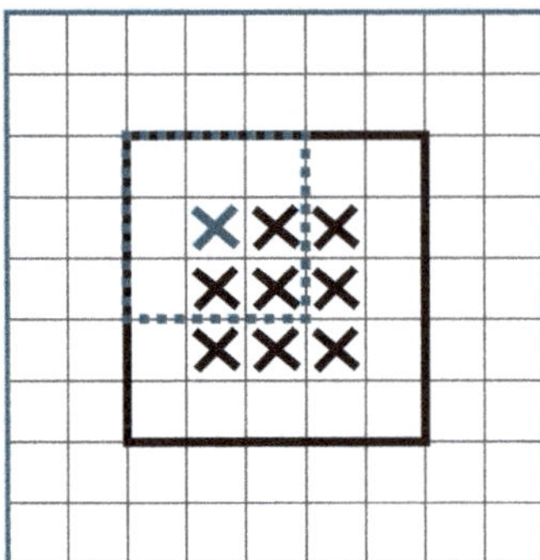

I looked at the work students had produced and saw that they had similar drawings for the 4 × 4 and 6 × 6 squares (see Figure B.4).

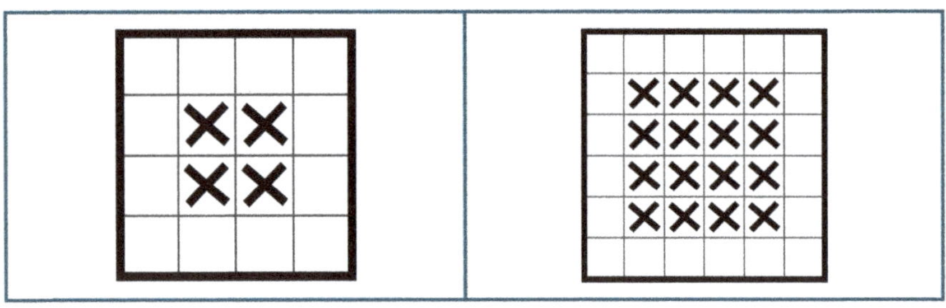

I then said, "OK, that now makes sense. But when I came over here, you were all staring at your paper. What were you doing?" Pasma replied, "Well, we were trying to figure out how
90   we were going to use this counting method to get the answer for a 60 × 60 square but we don't want to draw a 60 × 60 and move the 3 × 3 around." I said, "If you look across the squares that you have drawn, what do you notice about the shape that the Xs form in each square?" They paused and examined their work. "Oh," Sean said, "they make squares too." I then suggested that for each of the bigger squares they had completed, that they look at
95   the size of the bigger square and compare it to the size of the squares that were made by the center Xs, and see if there was a pattern they could identify. Then I left the group to work.

    WE REASON & WE PROVE FOR ALL MATHEMATICS

As I circulated through the room again, I stopped at different tables to ask about their progress and made notes on my clipboard about how different groups had approached the problem. If the students had found an answer to the 60 × 60 square question, then I

100 suggested that they start a poster to show their answer and how they arrived at it. At other tables, I checked to see that the students were working productively—that they had developed a systematic way of counting squares and then left them to work.

At this point, I noticed that there was only 10 minutes left in the class. I could see that a

105 number of groups had started to make posters of their solutions while other groups were still finishing up their work on the problem. After consulting with Gavin, we decided that we would let students finish up what they were working on and start tomorrow's lesson with a whole-class discussion of the Squares Problem. Since we had planned for a 2-day lesson, this seemed like a good stopping place for the day.

110

After class, I gathered up the posters that the students had completed. As I was reading the posters, I saw that most of the posters displayed counting techniques similar to one produced by Group 1, where students started with a 3 × 3 square in a corner, then systematically moved the square vertically or horizontally and counted (see Figure B.5). Justin,

115 Sean, and Pasma in Group 3 had been the only group to organize their counting in a little different way (see Figure B.6). I decided that I wanted to start the next day with a 10-minute discussion about these two solutions.

FIGURE B.5 Poster produced by students in Group 1.

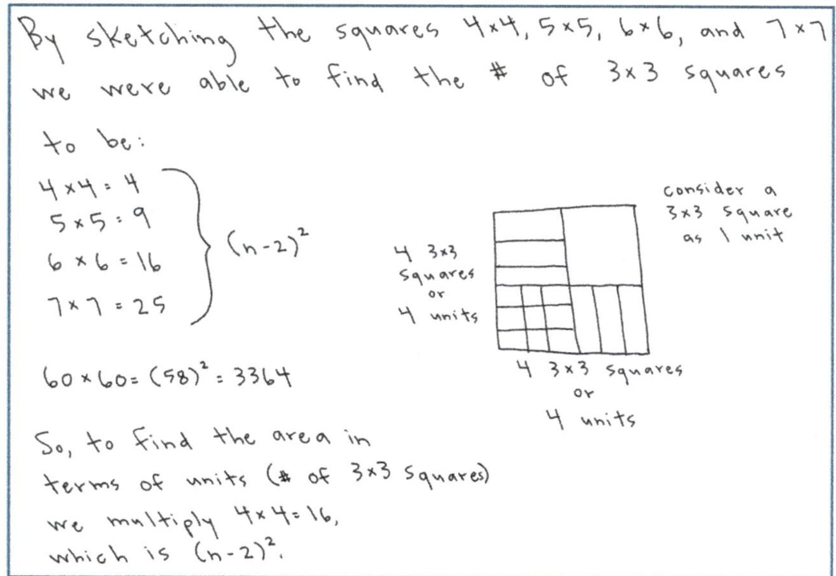

4x4 sq.
↓
2x2 sq.

5x5 sq.
↓
3x3 sq.

6x6 sq.
↓
4x4 sq.

7x7 sq.
↓
5x5 sq.

Original Square	Inside x Square
4x4	2x2 = 4
5x5	3x3 = 9
6x6	4x4 = 16
7x7	5x5 = 25
So	
60x60	58x58 = 3369

FACILITATING WHOLE-GROUP DISCUSSION ABOUT THE SQUARES PROBLEM (DAY 2)

120 As the students filed into class, I indicated to the students in Group 1 and Group 3 that we were going to start the class with them explaining their work from yesterday. Once the bell rang, the two groups presented their posters, and I orchestrated a discussion of the similarities and differences between the counting strategies used. I wanted to make sure that the class could see these two different counting strategies and understood how the two groups

125 arrived at their answers for the 4×4 through 7×7 squares. I made a table on a side whiteboard of the number of 3×3 squares in each of the larger squares as a point of reference for students (see Figure B.7).

 WE REASON & WE PROVE FOR ALL MATHEMATICS

Size of larger square	Number of 3 × 3 squares
4 × 4	4
5 × 5	9
6 × 6	16
7 × 7	25
60 × 60	?

As I turned around from the whiteboard, I stepped back and asked the students to look for patterns in the numbers. After a couple of minutes of think time, Bart said, "Hey, all of the
130 answers are square numbers." I asked the students to turn to a neighbor and discuss what Bart noticed and decide if they agreed or disagreed. Then I asked for thumbs up (agreement) or thumbs down (disagreement) and got all thumbs up.

Then Keisha called out, "Oh, the number of 3 × 3s is 2 less squared!" I asked Keisha to
135 explain what she meant by "2 less squared." She went on to say, "The number of 3 × 3s in the bigger square is two less than the side of the bigger square squared. So, the 8 × 8 square would have six squared—36—3 × 3 squares. Jillian added, "I think she is right." I gave the students a little bit of time to explore this conjecture, and by the end of this sharing time, there was consensus across the class that the "right" answer for the 60 × 60 square was 58 × 58 or 3,364.
140 At this point, I asked the students to respond in their notebooks to the following prompt, which I had prepared on a PowerPoint slide: How convinced are you that 58 × 58 or 3,364 is the right answer for how many 3 × 3 squares there are in a 60 × 60 square? Why are you convinced or not convinced? After giving them a couple of minutes to write, I asked them to talk about what they wrote at their tables for 2 minutes.
145
"So, how convinced are you?" I asked the class after 2 minutes. I heard comments like, "Pretty convinced," "I'm sure," and "Absolutely." I then asked, "Why are you so convinced that this is the right answer?" Maggie raised her hand, "I was convinced because we saw the pattern and we got the same answer, so it must be right." I saw students nod. Sam said, "I guess we
150 could really find out if we drew a 60 × 60 square and counted." I agreed that this would be very convincing and told the students that we were going to return to the Squares Problem

later after considering two other problems. I began passing out papers with circles printed on them in anticipation of starting the Circle and Spots Problem, the second task in the sequence.

INTRODUCING THE CIRCLE AND SPOTS PROBLEM

I launched the Circle and Spots Problem and encouraged the students to split up the work in their groups because I could see that we were running a little bit behind and I wanted to get to the punch line before the end of class. As I walked around, checking in on groups, I could see that students were using the sheets of circles and rulers to draw the pictures and count the non-overlapping regions using different numbers of spots. Some students were organizing their data in a table of values, but others were not. For those that weren't, I suggested that a table of values might be helpful to them for examining the data later. Once I was sure that we would have enough data, I called the class back together and began to construct a table of values on the board.

ORCHESTRATING WHOLE-GROUP DISCUSSION OF THE CIRCLE AND SPOTS PROBLEM

I started by saying, "You can fill in your table of values as we go along if you need to." Then I began to collect data on the board. "How many regions for one spot?" "One." "How many for two spots?" "Two." "How many for three spots?" "Four." "How many for four spots?" "Eight." I paused. "Let's take a minute and look at some patterns that might be emerging. What patterns do you see?" (See Figure B.8 for a version of the table at this point in the lesson.) After giving them a minute to think, I asked for volunteers.

FIGURE B.8 Table of values for the Circle and Spots Problem.

Number of Spots	1	2	3	4	5	6	
Number of Non-Overlapping Regions	1	2	4	8			

Stuart raised his hand. "I see that each time you add a spot, the number of regions doubles." I heard agreeing murmurs around the room. "Who agrees with Stuart?" A number of hands shot up. "Who disagrees with Stuart?" I looked around the room and saw no hands, but noticed that Rena and Dillon were pointing at a paper on their table and whispering. They had puzzled looks on their faces. "Rena," I said, "do you have a question to ask?" Rena looked up and said, "Well, our group got pretty far with our table of values, but it looks like we did something wrong, so we were trying to figure out what we did wrong." I asked them to explain. Dillon continued, "For five spots, we got 16 regions, which follows the pattern of doubling." I stopped him from going on and asked the class if they also got 16 regions for five

spots. They replied, "Yes." I asked Dillon to continue. "But when we counted for six spots,
we got 30 regions, which is not 16 times 2 or 32. So, I guess we didn't draw something right."
185 I moved to the chart and filled in 16 for five spots and 30 for six spots. I asked, "Who else
got 30 for six spots?" A few hands went up and I could see students going back to their
papers and talking to each other about something. I asked, "What's going on?" Rhonda
spoke first. "We didn't get 30 or 32 when we counted. We got 31." I went back to the table
of values and wrote 31 next to the 30. I turned back to the class and said, "Hmm. Can we
190 have two different answers for six spots?" I saw a lot of heads shaking no. "What could be
happening here?" I asked. Paula said, "Maybe somebody didn't draw very straight lines."
"That could be," I replied. "What else could explain the different answers?" Addy called
out, "Maybe we just counted wrong." Based on Addy's suggestion, I asked students to
recount the number of non-overlapping regions they got for six spots.

195

While they were busy recounting, I brought up a PowerPoint slide (see Figure B.9) with two
different versions of a circle with six spots and said, "Let's look at these two pictures, which
have straight lines. Both of these circles have six spots, but if you look closely, one has 30
regions and the other has 31. What's going on? Take a minute and look at these two circles
200 and talk at your tables about what you see."

FIGURE B.9 Two versions of a circle with six spots.

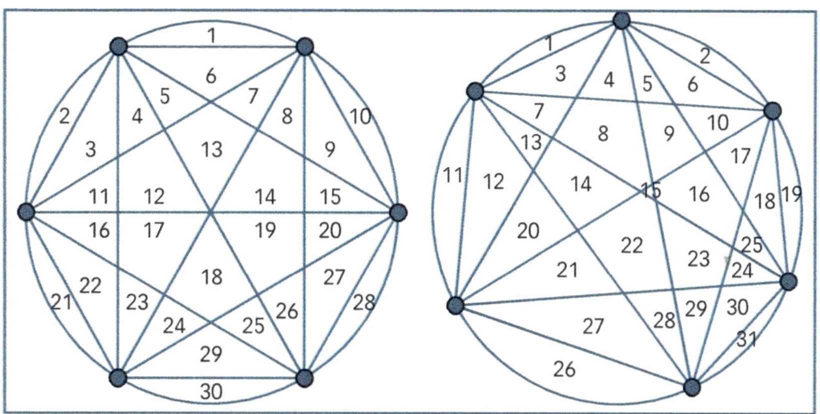

After a couple of minutes, I called the class back together. Dillon raised his hand. "We think
that it depends on where you put the spots." I said, "Talk about that a little bit more." He
continued, "Well, on the left, when the spots are pretty even around the circle, the lines meet
in the middle. But in the circle on the right, the spots aren't even and the lines don't meet in
205 the middle." I asked the rest of the students to use thumbs up or thumbs down to show if
they agreed with Dillon or not. I saw thumbs up. "So, if you agree with Dillon, that there

is a way that we can get two different answers for six spots, what does that mean about the pattern that we noticed going on based on one spot through five spots?" I called on Shonda, who said, "I guess we can't say that it doubles. The pattern we saw in the first five doesn't continue." "So," I said, "could you tell me with confidence how many non-overlapping regions there would be for 15 spots?" Students shook their heads no.

"Well, then," I said. "I'm a little confused. On the Squares Problem, you told me you were convinced that you had the right answer for a 60×60 square, even though we didn't actually draw one. We based our answer for a 60×60 square on the pattern we saw emerging in the first few cases. But now you are telling me that we can't trust a pattern that we see emerging?" I let that sink in for a minute. "At your tables, talk about what you've learned from doing the Squares Problem and the Circle and Spots Problem."

After a few minutes, I said, "What I heard you talking about as I walked around is that many of you felt like you couldn't trust a pattern from just a few examples and you are doubting the answer we got to the Squares Problem. Would a hundred examples be more convincing?" I could see some heads nod yes and some heads shake no. "A thousand?" Again, some nods and some head shakes. "Not sure?" I asked. "Let's consider one more problem."

INTRODUCING THE STORY OF THE MONSTROUS COUNTEREXAMPLE

I advanced to the slide with the Story of the Monstrous Counterexample (see Figure B.10) and asked for a volunteer to read it aloud while everyone else read along silently. Katie did a good job of reading aloud to her classmates, stumbling only when she had to read the final number in the story. She said, "I don't even know how to read a number that big!" I smiled and said that it was OK, that everyone could see the number.

FIGURE B.10 The story of the Monstrous Counterexample.

A group of mathematicians was exploring outputs for the expression $1 + 1141n^2$ when n is a natural number. After evaluating the expression for several natural numbers and looking for any patterns that seemed to be emerging, they made the conjecture that this expression *never* returns a square number. To test their conjecture, they wrote a computer program to evaluate the expression for all natural number inputs and to stop when the output tested as a perfect square. Then they left the computer to its work.

They kept checking in with the computer's work, and as n grew bigger and bigger without an output of a square number, they became more and more confident that their conjecture was true. Imagine their surprise when they returned to the computer and found that it had stopped running the program, meaning that the program had an output of a square number. They were even more surprised to find that this expression does return a square number, but not until $n = 30,693,385,322,765,657,197,397,208$.

    WE REASON & WE PROVE FOR ALL MATHEMATICS

"Look at how many examples the computer had to check before it found a number that made this rule true," I said. I heard students comment, "Wow," "That's cool," "That's a really big number," and "No way." "So," I continued, "can we really know how many
235 3 × 3 squares are in a 60 × 60 square based on a pattern we saw with smaller squares?" There were murmurs of no and some shaking of heads, but I wondered how many students really believed it.

As class was coming to a close, I handed out the exit slip, which said, "What did you
240 learn today about using examples to make a valid mathematical claim or argument?" As I watched the students write their answers, I hoped that when I read these slips later that afternoon, they would tell me that they had begun to understand that there has to be something more to making valid mathematical arguments than just using examples. Once we establish that examples are not enough, then we will be ready to pursue what *is* needed to
245 construct a valid argument and we could revisit the Squares Problem to write a valid argument that proves the conjecture that the number of 3 × 3 squares in a $n \times n$ square is $(n-2)^2$.

Appendix C

Writing and Critiquing Proofs: The Case of Vicky Mansfield

For the past 3 days, Vicky Mansfield and her high school students have been working on the first unit in their mathematics textbook, which introduces flowchart proofs. Vicky really likes the flowchart proofs, even though they are very different from the two-column proofs that she produced as a student, because the flowchart format makes it possible to clearly
5 *follow her students' line of reasoning. In the lesson featured in the case, Vicky's students are working on writing flowchart proofs that integrate the intersecting line theorems that they just learned with previously learned definitions, theorems, and properties. During this lesson, Vicky wants her students to learn that*

10   1. *definitions, postulates, and theorems can be used in proving conjectures;*

   2. *conjectures can be proven in different ways; and*

   3. *critiquing the reasoning of others can result in greater understanding about the math-*
15   *ematical relationships as well as the clarity and completeness of the proof under discussion.*

Vicky hoped that she would observe some variety in her students' proofs. In the past 2 weeks since school started, she had noticed that her students have been very concerned
20 *about solving problems "the right way"—meaning **her** way. Vicky wants her students to think for themselves and solve problems in ways that make sense to them, so she wants to address this concern as soon as she can.*

VICKY TALKS ABOUT HER PLANS
25 It is the end of the second week of school, and today we are working on one of the "Summarize the Mathematics" tasks from our book. These tasks are intended to pull together the ideas that students have explored up to that point. Although I expect that in proving their conjectures students will draw upon the definition of perpendicular, the Linear Pair Postulate, and the Vertical Angle Theorem that we have recently discussed, I am hoping that they will come
30 up with a variety of different ways to prove their conjectures. I don't want students thinking that there is only one way to solve a problem or to prove a conjecture. If I see some variety in the proofs students produce, then I plan to use this as an opportunity to have students critique different arguments. Critiquing arguments is an important skill, but not something we have done yet, and I want students to get experience in thinking through and critically
35 analyzing solutions they didn't create.

I started class by getting students' attention and asking everyone to take a look at the diagram that I had drawn on the whiteboard (shown in Figure C.1). The diagram came from the textbook task that we were going to work on today. I didn't want to present the entire task by referring them to their textbook quite yet, because in the textbook the diagram included additional information (that two of the angles were equal) that I didn't want to share immediately. I had labeled each of the angles in the diagram with a number (e.g., $\angle DBA$ is $\angle 1$), just to make it a little bit easier for students to write and talk about the angles. I asked students to take out their notebooks, copy the diagram, and spend a few minutes on their own making as many observations as they could about the diagram. I then added, "I'm not looking for a particular number of observations—I just want you to be

FIGURE C.1 The diagram that students were given.

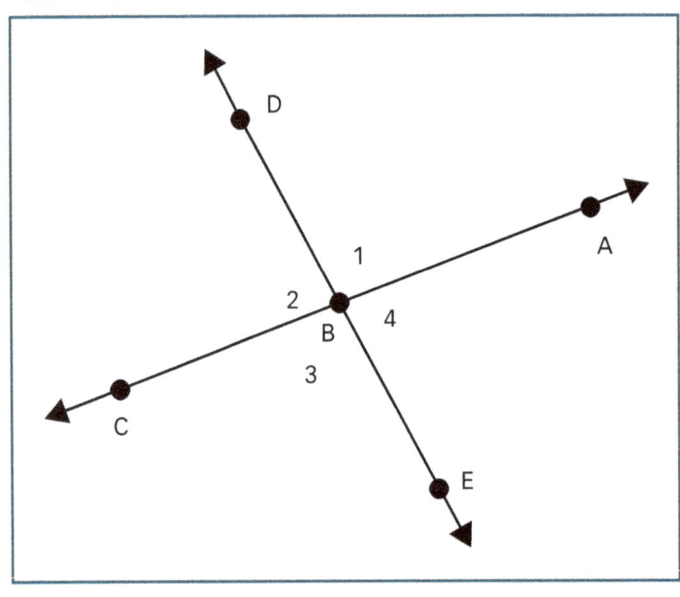

thinking and writing for the entire time I give you." I felt like I needed to add this because earlier this week, some students had stopped working as soon as they had one idea or solution written down, and I wanted to make sure students knew that this was not acceptable.

As students worked, I walked around the room, reading their observations and making some notes on my lesson plan. After a few minutes, I asked students to stop writing and to turn to the "Summarize the Mathematics" task on page 34 of their textbook (shown in Figure C.2). I asked them to see what conclusions they could come to about the angles in the

FIGURE C.2 The problem that was the focus of the lesson.

Summarize the Mathematics

In this investigation, you used deductive reasoning to establish relationships between pairs of angles formed by two intersecting lines. In the diagram, suppose the lines intersect so that $m\angle DBA = m\angle CBD$.

a. What can you conclude about these two angles? Prepare an argument to prove your conjecture.

b. What can you conclude about the other angles in the diagram? Write a proof of your conclusion.

c. What mathematical facts did you use to help prove your statements in Parts a and b? Were these facts definitions, postulates, or theorems?

d. Describe the relationship between $\overleftrightarrow{AC}$ and $\overrightarrow{DE}$.

Be prepared to share your conjectures and explain your proofs.

Source: Fey, J. T. et al., (2009)

55   diagram (Parts a and b), given the additional information that was provided. I suggested that the observations they had just made might be helpful if they weren't sure how to get started.

After a few minutes, I told students that I wanted them to work with their groups to make a conjecture about the relationship between lines AC and DE (Part d) and then create a flowchart

60   proof that would prove their conjecture. I thought the questions in the book were a little confusing since Part a asked for an argument and Part b asked for a proof. I didn't think students needed to create two different proofs or arguments, so I settled on something a little different. I reminded students that the bulletin board in the front of the room described my expectations for group work (shown in Figure C.3) and suggested that they refer to it as needed.

FIGURE C.3 Vicky's expectations for group work.

Groupwork

All group members …
- begin work promptly
- face each other
- work on each problem together, at the same time
- ask questions, provide ideas, and listen to each other
- stay on task
- are respectful and encouraging, and involve other members

 MONITORING SMALL GROUP WORK

As students discussed the task in their groups, I walked around the room and occasionally pulled up a chair to a group so that I could listen more carefully. I noticed that the students in Group 3 (Quentin, Elisa, Becky, and Rob) didn't seem to be talking to each other. I made my way over to them, and Elisa said to me, "Ms. Mansfield, we're stuck." I had a hard time believing that the four of them couldn't make any progress together. I took a quick glance at their observation lists and I noted that each of them had recorded several observations about the diagram that I knew would be helpful in proving that line AC was perpendicular to line DE. I asked, "Well, have you talked to each other? What kinds of things can you say about the diagram?" Rob replied, "All we know is what the problem tells us, that angle 1 equals angle 2." I responded by saying, "OK. Is there anything else that you know about angles 1 and 2? Go back to your observation lists." Becky said she thought that "angles 1 and 2 would add up to 180 degrees," and Elisa said that she had made that observation, too. Quentin then joined in the discussion, pointing out that "the only way that angles 1 and 2 can be 180 degrees *and* be equal is if they're both 90." Elisa excitedly added, "Which would make the lines perpendicular!" I told them that they had made some interesting points, and now they needed to figure out how to arrange their ideas into a flowchart proof that others would be able to understand. As I got up to leave their table, I added, "See how much progress you can make if you just talk to each other?"

As I continued walking around the room, I noticed that while most of the seven groups were writing proofs that drew upon the reasoning that Group 3 had just described, Group 5 (Jason, Stacie, Kendall, and Tim) seemed to be creating a different sort of argument. I sat down with them and looked at what they had recorded so far (shown in Figure C.4). I asked, "So is your proof complete? Have you provided enough so that even a skeptic would be convinced?" Tim suggested that some people might wonder how they knew that all the angles added up to 360 degrees. I underscored his idea by asking, "Yes, that's a good question. How *do* you know it's 360? Is there anything you can add to your proof that would explain where the 360 comes from?" Kendall looked through her observation list and her notes from earlier in the week, and said to me, "What about the Linear Pair Postulate?" I suggested that she ask her group what they thought about this, and left to monitor the work of the other groups.

FIGURE C.4 The proof that Group 5 produced during small group work.

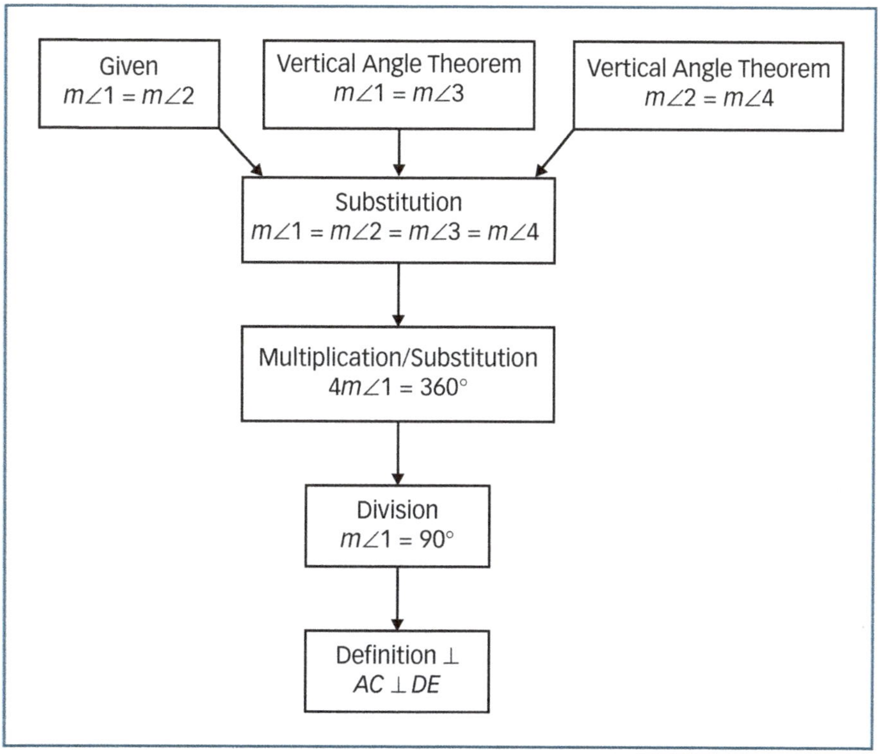

FACILITATING WHOLE-GROUP DISCUSSION

We had about 30 minutes left, so I wanted to start the whole-class discussion. I decided to ask Group 1 and Group 5 to record their proofs on the whiteboard while the other groups finished writing their proofs. Group 1's proof was pretty similar to the proof written by Group 3 (and all the other groups in the class), so I wanted to start with that one. Group 5's proof was really different from everyone else's, so I wanted to save it for last.

I started the discussion by focusing everyone's attention on Group 1's proof (shown in Figure C.5) and said, "Let's be critical—in a respectful way. Do you see anything that you don't understand? Do you see places where further explanation is needed?" Since it's so early in the year, I felt that I needed to ask more specific questions than simply asking my

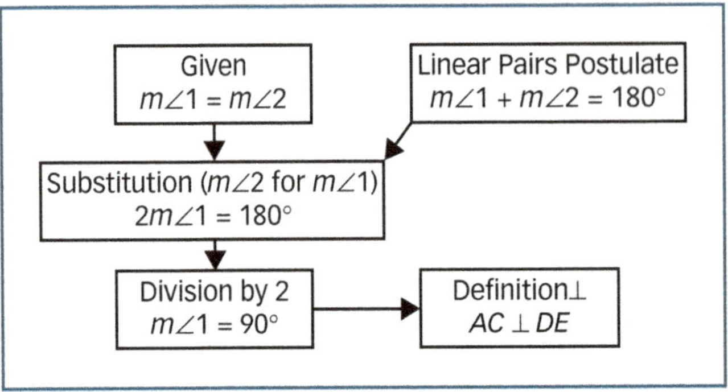

typical "Does anyone have any questions or comments about this?" question that I usually ask after a student or a group shares a solution. I've noticed that sometimes students just don't know how to respond to this admittedly vague question! I've found that being more specific, especially at the beginning of the year, helps students understand what I want them to focus on.

I waited for a few seconds, and no one said anything or raised a hand. I tried to encourage everyone by adding, "Remember, these were not put up here because I thought they were necessarily perfect, and it's not my opinion that matters anyway." I waited and watched the second hand on the clock in the back of the room move 20 painfully long seconds. I suppose that I could have gotten the discussion going myself by asking Group 1 some questions, but I didn't want students to get used to me taking over things if they got stuck! But I just couldn't take the silence anymore, so I asked, "Any questions?" I was relieved to see Katie raise her hand, and I quickly acknowledged her. She started by saying, "Since we're substituting the measure of angle 1 for the measure of angle 2, you could say the measure of angle 1 plus the measure of angle 1, and then just change it to two times the measure of angle 1 equals 180." I asked Katie to come to the board and add her idea to the proof (shown in Figure C.6).

I thought this was an interesting addition, but it seemed incomplete to me. I asked, "Does she need to say *what* she substituted?" I heard a chorus of "mhm's" from the class as Katie replied, "Oh, yeah," and added that she was substituting $m\angle 1$ for $m\angle 2$ to the proof on the board. Katie continued her idea by making a few more revisions to the proof, as shown in Figure C.6. She explained, "So now the reason for the two times the measure of angle 1 equals 180 is combining like terms instead of substitution, because, if you erase this arrow [from the $m\angle 1 + m\angle 2 = 180$ to the $2m\angle 1 = 180$ in Figure C.5], and add an arrow here [from the $m\angle 1 + m\angle 1 = 180$ to the $2m\angle 1 = 180$ in Figure C.6], that would be combining like terms to justify the $2m\angle 1 = 180$." Katie also replaced the text in the middle left box with "Combining like terms." I said, "OK. Does she need any more arrows?" Steven said, "I think she needs an arrow from the Given to the Substitution." As Katie added that arrow, I said, "Good. 'Cause if you're going to put angle 1 in for angle 2, you need to tell me that they are actually the same. Very good, thank you, Katie."

FIGURE C.6 Group 1's proof with Katie and Steven's additions shown in gray.

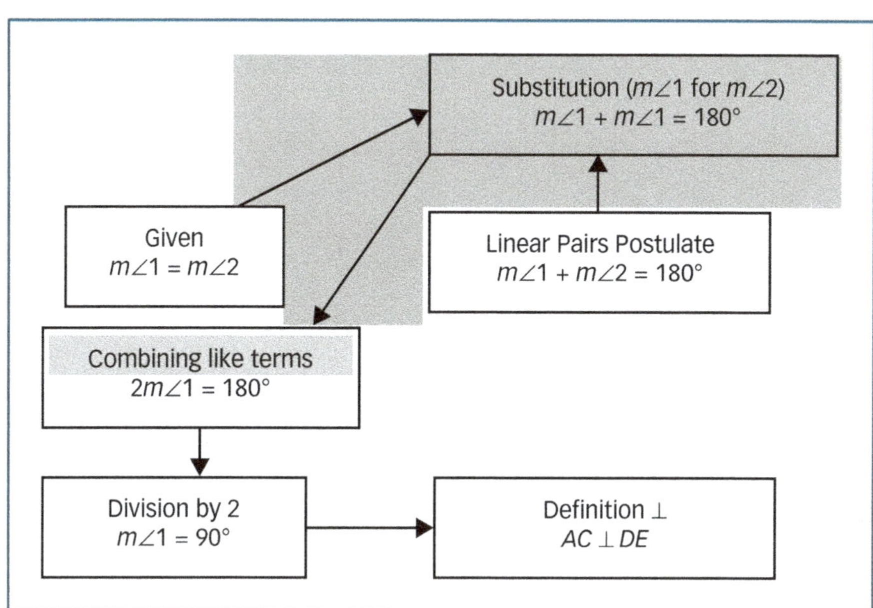

I was pleased that students had suggested some revisions to Group 1's proof. Since we only had about 20 minutes left, I wanted to move on to Group 5's proof (shown in Figure C.7). I thought everyone needed a chance to think about this proof on their own first, since it was so different from the way in which Group 1 (and all the other groups) wrote their proof. I said, "I want you to take a minute and digest this. If you have any questions or see anything that you don't think is clear, write them down in your notebook." I let almost an entire minute pass and then I asked, "Comments? Or are we still thinking?" No one asked for more

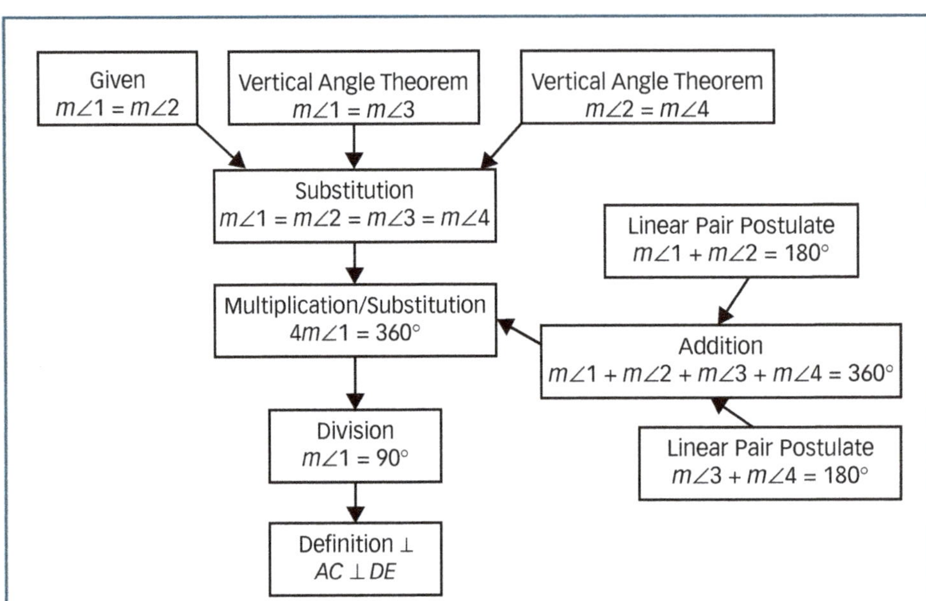

145   time, and I noticed that Tamiko had her hand raised. I called on her, and she commented, "Well, I just wanted to ask about the addition and the two Linear Pair Postulates parts of the proof. Could you do without those? You don't necessarily need those." She paused for a moment, and I waited for her to finish her thought. She added, "'Cause they are kind of their own thing out there. So are they really needed? Couldn't you just do it with the multi-

150   plication/substitution part going to the division part?"

Adam responded to Tamiko's wondering with another question, "But then how would we prove that the four angles add up to 360 degrees?" This was the same issue that had come up in my conversation with Group 5 during small group work, and I was glad that Adam

155   raised this point. I flagged this as an important issue for us to consider by saying, "So if we erased the parts about addition and the Linear Pair Postulate, then Adam is saying, the reader would wonder where the 360 degrees came from. Any ideas about that?"

Jackson, who has been a bit of a class clown so far this year, said, "Just 'cause it does!"

160   and the entire class responded with laughter. I smiled, but continued the discussion by ask- ing, "Is that going to be good enough for a skeptic? 'Cause it does?'" Will then suggested "proving by multiplication and then substitution," and Tamiko nodded and said that's what she had been thinking, too. I really wanted us to come back to Adam's point, about where

        WE REASON & WE PROVE FOR ALL MATHEMATICS

the 360 degrees comes from, so I asked, "So if we just left all this off—the addition and Linear Pair Postulates parts—what would my question about this be?" Several students responded with questions such as, "How'd you know it was 360?" and "Where'd the 360 come from?" I responded, "Right, I'd want to know how you knew those angles all added up to 360 degrees. So how would you respond to that?"

Stacie was waving her hand wildly, so I called on her, and we began the following exchange:

Stacie: Couldn't you just look at the picture? Like there are four angles total. And you know that four angles make 360 degrees around a point.
Me: How do I know that four angles equal 360 degrees?
Stacie: Like if we drew it on the picture, would we still have to write it in, in the proof?
Me: So is it okay for me to draw a circle on our original diagram and say, "Well, that means it must be 360 degrees." Is that your question?
Stacie: Well, no, not really.
Me: Okay, well, ask it again because I didn't understand it.
Stacie: Well, I don't know what to ask.
Me: Well, can you just repeat what you said before?
Stacie: Um, like, you know as we're going along, saying that the four angles all equal each other, like you have the first three and then the next one down, like while you're doing all of that, couldn't you just draw the marks on your picture and eventually it would show that they all equal each other? And you know that there are only four angles there because of the picture.
Me: Okay, so my challenge to you is still, do any four angles add up to 360 degrees?
Several students [muttering]: No.
Me: And can I trust this picture?

Paige seemed to understand what Stacie was saying because she chimed in, "So if the angles are formed by lines that intersect at a point, then they will always equal 360 degrees." Stacie responded, "Yeah, that's what I was thinking." Paige then added, "But if they're just any four angles, not necessarily around a point, then they could add up to anything, not just 360 degrees."

Stacie and Paige had just come up with an interesting conjecture that, to be honest, I hadn't thought about before. I wanted us to continue discussing this conjecture, but first I wanted to make sure we were all on the same page. I said, "Stacie and Paige have made a really interesting conjecture. Can someone repeat it in their own words?" No one raised a hand, so I said, "Group 3, we haven't heard from you in a while. Can you restate their idea in your own words?" The students in Group 3 looked at each other, waiting for someone to speak up. Finally, Quentin said, "If you have four angles around a point, then they have to equal 360?" I asked Stacie and Paige if that captured their idea, and they agreed.

I wondered if this conjecture would always hold up and tried to think of a situation in which
it wouldn't. Off the top of my head, I sketched a diagram on the whiteboard that might
press students to explore the conjecture, as shown in the left-hand diagram in Figure C.8. I

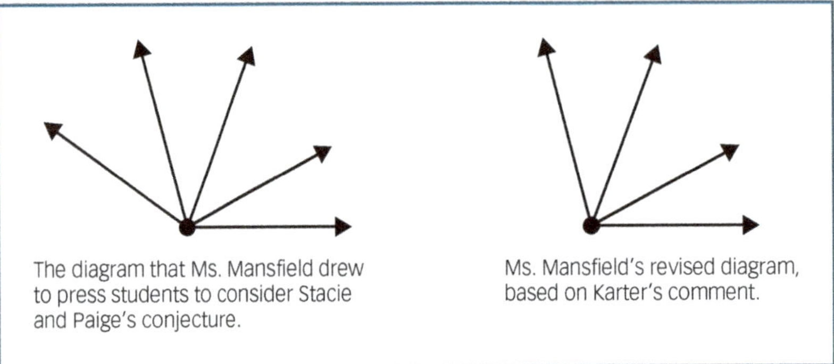

The diagram that Ms. Mansfield drew
to press students to consider Stacie
and Paige's conjecture.

Ms. Mansfield's revised diagram,
based on Karter's comment.

said, "Okay. I know this is not the same picture we've been working with today. But here are
four angles and they are all around a point. Do these four angles equal 360 degrees?" Karter
called out, "That shows five angles, not four. I think you're forgetting the outside one." I
made the adjustment Karter seemed to be suggesting (shown in the right-hand diagram in
Figure C.8) and asked, "Okay, so you're saying I need to erase one of these rays and count
the obtuse angle [pointing at the white space to the left of, and under, my diagram] as one
of my four angles?" Karter nodded his head in response to my question.

Since I couldn't seem to come up with a quick counterexample off the top of my head, I didn't
want to spend any more class time exploring Stacie and Paige's conjecture without being more
prepared myself. So I flagged this discussion as a work in progress by saying, "Okay, so this
weekend I want you to think more about this and see if you can come up with a counterexample.
Do four angles around a point *always* equal 360 degrees? Okay? And we'll come back to that."
We only had about *5* minutes left and I still had some questions about Group 5's proof
that hadn't been brought up yet. Since we were running out of time, I made a quick decision to focus on one of my questions. I pointed to the top section of the proof (shown
in Figure C.9), which focused on the claim that all four angles were equivalent. I asked,
"Is this enough to prove that all four of those angles truly are the same? How do I know
angle 1 equals angle 4?" Several students responded by noting that "1 equals 2, and 2 equals
4." Although I followed their line of thinking, I wasn't satisfied by this explanation. I asked,
"So I'm still not following how you know that angle 1 equals angle 4. Am I supposed to
read your minds?"

WE REASON & WE PROVE FOR ALL MATHEMATICS

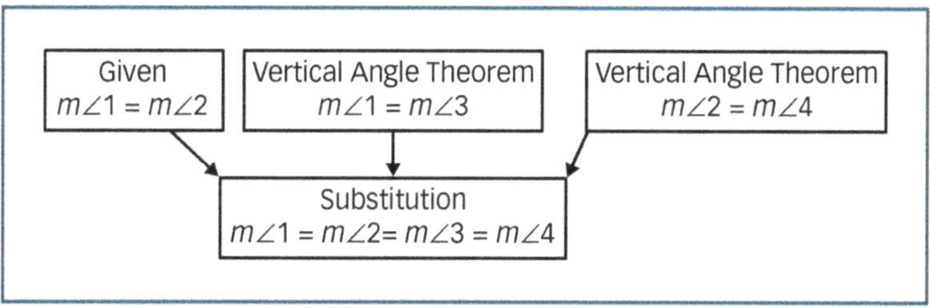

Emma added that we needed to add another piece about substitution. I asked if she could come to the board and show us, and she recorded "Substitution $m\angle 1 = m\angle 4$" (shown in Figure C.10), but didn't use any arrows to connect her new information to the original proof. I added, "We can't have floating ideas in our proof. I think we need some arrows here." Emma then inserted arrows from $m\angle 1 = m\angle 2$ and $m\angle 2 = m\angle 4$ to the Substitution piece that she had just added (shown in Figure C.10). I asked the class if Emma's additions

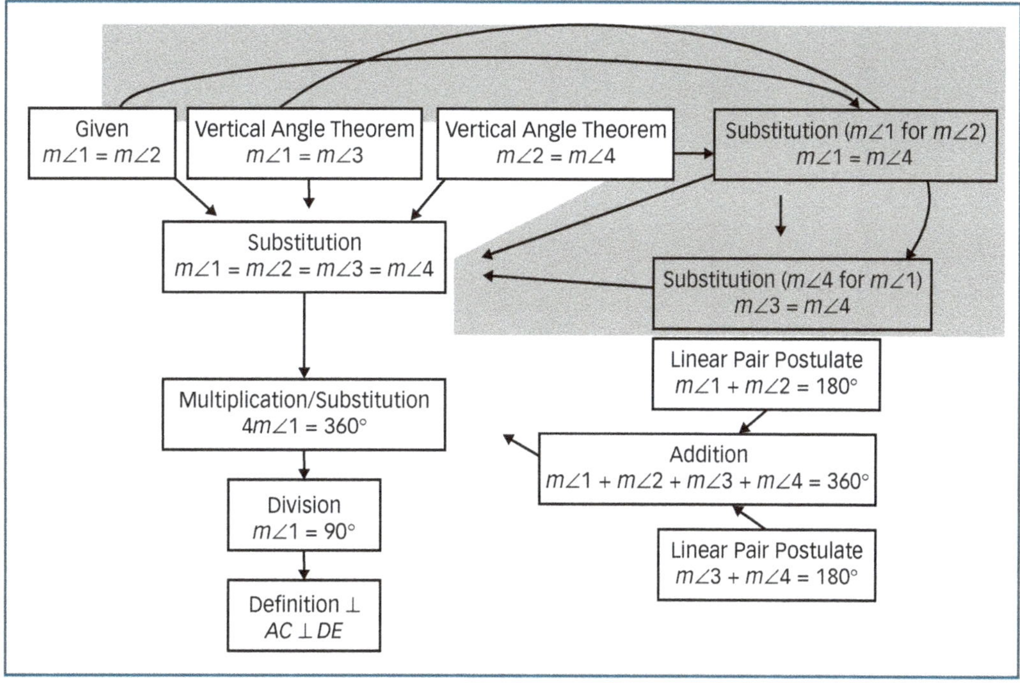

were enough to prove that all four angles were equivalent. Kevin said that he thought we should add something about angles 3 and 4 being equal. I asked him to come show us what he meant, and he began adding his idea to the proof, as shown in Figure C.10.

I added, "While Kevin is writing his idea on the board, I want you to write the question we were discussing earlier in your notebook, and to think about it over the weekend—Do four angles around a point always equal 360 degrees?" As students were recording this question in their notebooks, I asked, "Is it okay that some of us approached these proofs in different ways?" Students weren't sure if this was a rhetorical question, but many mumbled, "Yeah," as I was also saying, "Yes." I followed this up by asking, "Could we have added something to both of these proofs to make them a bit more clear? Absolutely. Is it okay that your group members don't think like you do?" Jackson was quick to jokingly say, "No!" I gave him a stern look and said, "Yes, it is okay."

I also noted to the class that I was leaving Group 5's proof on the whiteboard, because I wanted to continue our discussion about this on Monday. The bell rang, and students made their way out of the room.

WE REASON & WE PROVE FOR ALL MATHEMATICS

Appendix D

Pressing Students to Prove It: The Case of Nancy Edwards

In a professional development session that Nancy Edwards attended at the beginning of the year, she was reminded of the importance of asking students to explain why things work the way they do and how they know they will always work. During the session, Nancy and her colleagues from high schools around the city were asked to compare a task from the textbook that was currently being used in the district (Task A1 in Figure D.1) to a modified version of the task (Task A2 in Figure D.1). Nancy and her colleagues were amazed that such a minor change to a task could have such a major impact on the opportunities it could provide to students. The teachers were in agreement that while both tasks asked students to make conjectures, Task A1 did not require students to think about why the conjecture was true or how they knew that it would always work. Furthermore, Task A1 could leave students with the impression that finding a pattern and making a conjecture are sufficient to convince someone that something is always true.

FIGURE D.1 The original textbook task and its adaptation.

TASK A1	**TASK A2**
MAKING CONJECTURES—Complete the conjecture based on the pattern you observe in the specific cases.	Complete the conjectures below based on the pattern you observe in the examples. Then explain why the conjecture is always true *or* show a case in which it is not true.
Conjecture: The sum of any two odd numbers is: ________. $1 + 1 = 2$　　　$7 + 13 = 20$ $1 + 5 = 6$　　　$15 + 19 = 34$ $3 + 5 = 8$　　　$201 + 305 = 506$	**Conjecture:** The sum of any two odd numbers is: ________. $1 + 1 = 2$　　　$7 + 13 = 20$ $1 + 5 = 6$　　　$15 + 19 = 34$ $3 + 5 = 8$　　　$201 + 305 = 506$
Conjecture: The product of any two odd numbers is: ______. $1 \times 1 = 1$　　　$7 \times 13 = 91$ $1 \times 5 = 5$　　　$15 \times 19 = 285$ $3 \times 5 = 15$　　　$201 \times 305 = 61{,}305$	**Conjecture:** The product of any two odd numbers is: ______. $1 \times 1 = 1$　　　$7 \times 13 = 91$ $1 \times 5 = 5$　　　$15 \times 19 = 285$ $3 \times 5 = 15$　　　$201 \times 305 = 61{,}305$

Source: Adapted from Larson, R. et al. (2004).

In planning for the first week of school, Nancy decided that she would implement the modified version of the task (A2). She thought that a positive feature of the task is that the argument for why the conjecture holds true for all cases could be made in many different ways. Although the task is a number theory task rather than a geometry task, she wanted to start the year with a task that students would have access to—that is, they were not learning new content at the same time they were learning new processes (reasoning-and-proving). As a result of engaging in this task, she wanted her students to learn the following:

1. *For a mathematical argument to be a proof, it must show that the conjecture is true for all cases.*
2. *Proofs can utilize different representations, and those representations can be connected.*
3. *Examples alone do not constitute a proof.*

NANCY TALKS ABOUT HER PLANS

This year, I am determined to do more in my geometry class than I did in my first year of teaching. I feel like I spent too much time telling students how to do procedures and making them practice, and I wanted to start implementing more tasks that require my students to think and justify what is happening mathematically in the tasks. While I still want my students to be able to perform routine procedures accurately and to quickly recall key definitions and theorems, I also want them to be able to make sense of the mathematics they are learning. So, after attending the professional development session this past summer, I decided that proving things would be an ongoing feature of the work in my classes. I think that students working on Task A2 is a great way to dive into this kind of thinking. In planning for the task, I made copies for the students and made sure that there were manipulatives (multilink cubes and square tiles) available for them to use if they wanted to.

THE CLASS BEGINS

Students entered the classroom weighted down with their backpacks and chatting with their friends. They took their seats in their assigned groups of four and pulled out their homework from the previous night, their notebooks, textbooks, and pencils—all part of the daily routine to maximize every minute. When the bell rang to begin class, I asked my students, "What does it mean to prove?" I gave them a few minutes to think about this privately and then asked if anyone had any thoughts they wanted to share. Several hands shot up and I asked Miray to get us started. She said, "It means to give evidence to support what you are saying. To persuade someone." I asked the class, "How much evidence do you need? What has to happen in a courtroom to convince a jury?" Hands shot up around the room and I called on Devis, who responded, "You need to have enough evidence to be convinced beyond a reasonable doubt." I asked the class what this meant. Renee volunteered, "You have to show that something *has* to be true." I asked, "How can you show that something is true all the time?"

   WE REASON & WE PROVE FOR ALL MATHEMATICS

At this point, I left the question unanswered and passed out the modified version of the text-book task that had been presented in the professional development session (Task A2 as shown in Figure D.1). I told the students that we were first going to focus on the sum of two odd numbers part—depending on how far we got in class with this part, I knew that we could go to the second part at the end of class and students could finish their arguments for homework. I asked the students to silently examine the examples and paused for a few seconds. Then I asked how they would fill in the blank and they replied in unison, "Even." I said, "How do you know? Can you prove that it always works?" Philip said, "Look, Mrs. E, we have a few examples here and we can see that it will always be true. If someone is not convinced, we could check more examples and I bet you that the statement will be true no matter how many we try." I replied, "Are examples enough to prove that something is always true?" Chris said, "In this case, examples are good enough for me. It would probably even convince a jury." I chuckled but held firm. "Well, maybe so, but it is not good enough to convince skeptics like me."

I told students that their job was to show that when you add any two odd numbers together, you always get an even number. I told them they could use words, pictures, numbers, or anything else they could think of in creating their argument—to explain why the conjecture is always true or to show a case of when it is not true (following the directions of the task). I told them they would have 25 minutes to develop an argument and to create a group poster. They know where to find the poster paper, markers, and any other tools they might need. I set the timer for 25 minutes so that we could all keep track of the time.[1] I have found that I need to keep students accountable for producing a product within a specified amount of time. Otherwise, we spend too much time on one problem, and not always productively. I also lose track of time and 25 minutes turns into 40!

MONITORING SMALL GROUP WORK

As the groups got started on the task, I grabbed a notepad from my desk so I could keep track of the strategies that different groups were using and make note of things I wanted to be sure to bring out in the whole-class discussion. The first thing I wanted to do was just take a quick walk around the room and make sure that everyone was off to a productive start. I noticed that all of the groups were working on justifying that the conjecture is always true—no one seemed to be trying to find a counterexample. I made a note of this on my pad.

What I also saw during this quick walk around the room is that most groups were trying more examples—almost as if they thought that showing more examples would be convincing that the conjecture holds for all cases. While examples can be useful in seeing patterns and making conjectures, the task already provided many examples, and students were

---

[1] I have been using the web-based timer that can be found at http://www.timeme.com/timer-stopwatch.htm. This is much better than my old-fashioned timer since it is really big, and if I turn my computer to face the class, everyone can see it!

90 already convinced that this conjecture worked for all cases. I decided to interrupt the group
work for a minute. I called the class together and said, "Examples may be a reasonable place
to start, but remember, you will need more than examples to show that the conjecture will
always be true. Think about what you know about odd and even numbers."

95 As groups resumed their work, I saw that Group 6 had already started making their poster.
Surprised that they could be at this point already, I decided to take a closer look. I asked
Nina, one of the members of the group, what they had decided. She explained, "We decided
to try a few more examples (see Figure D.2) and to include negative integers as well as positive ones to make sure it still worked and it does." I was impressed that they had considered
100 whether the conjecture would hold true for all integers, but was hoping there was more to
their argument. I asked the group how they could convince a skeptic that this would always
work and that there wasn't one that didn't work that they just hadn't found yet. They said
they weren't sure where to go from here. I asked, "What do you know about odd numbers?"
Tamika indicated that when you divided them by two there was one left over. I said that
105 this was an interesting idea and suggested that they try to build a model or draw a picture
to show this. I told them that I would come back in a few minutes and check in with them.

FIGURE D.2 Work produced by students in Group 6.

The sum of two odd numbers equals an even.
$1 + 1 = 2$
$3 + 3 = 6$
$5 + 3 = 8$
$5 + 5 = 10$
$-1 + 3 = 2$
$7 + 7 = 14$
$-21 + -11 = -32$
Group 6: Devis, Nina, Tamika, and Mike

At this point, I wanted to make sure that the remaining groups were making good progress
and to see which group might need the most help. Here is what I found:

110 - Group 3 had taken an algebraic approach and had represented the two odd numbers
  as $2x + 1$ and $2a + 1$.
  - Group 5 did something similar, but instead of using $2x$ to represent an even number,
  they simply used the word even and then represented an odd number as even + 1.

- Group 2 had created a narrative argument that was built on the idea that every odd number had a loner number and that if you add the two loner numbers together it equaled an even number. While their argument was general, I wanted to make sure at some point that they were clear about this notion of loner numbers.

I thought that these groups were okay for now and turned my attention to Groups 1 and 4.

120 Group 1 was drawing a diagram on their poster, as shown in Figure D.3. I asked them to explain what they were doing. Angela volunteered that they decided to show that when you added 11 and 13 you really had five sets of 2 for the 11 with one left over and six sets of 2 for the 13 with one left over, and if you add the two leftover ones, you got another group of 2. I asked them whether this proved that the sum of any two odd numbers was always even.

125 Chris said, "Well, we showed one example. Now we have to figure how to show it for all of them without drawing more pictures."

FIGURE D.3 Work produced by students in Group 1.

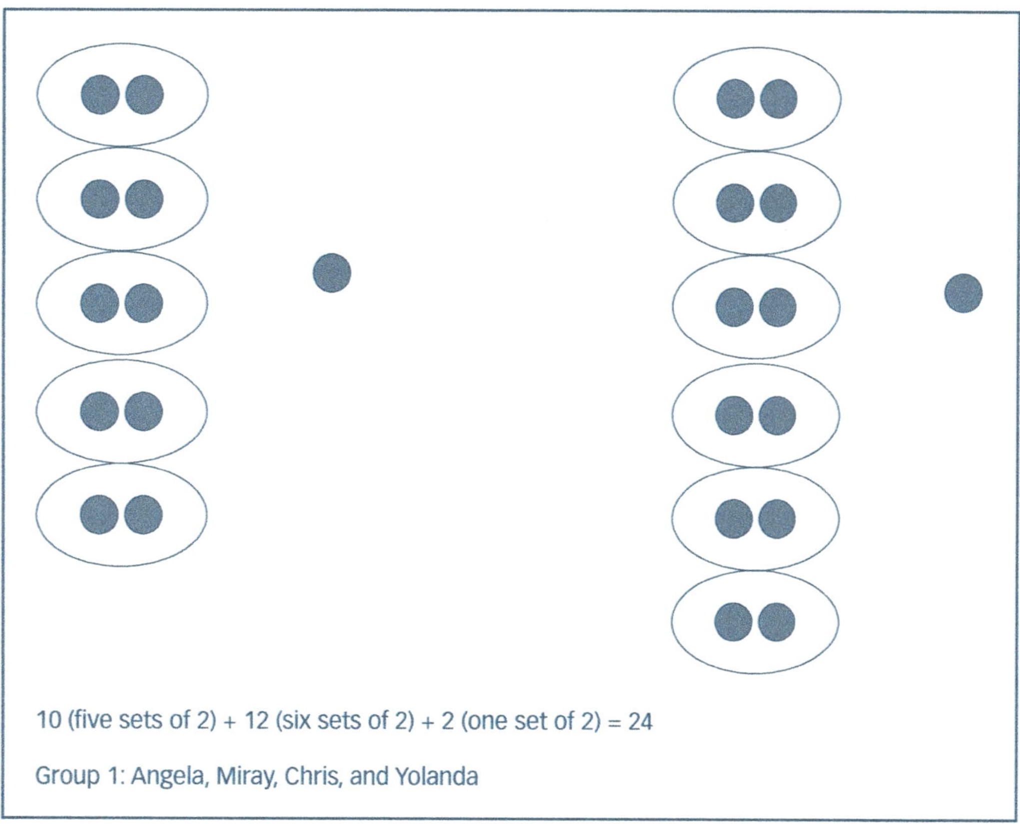

At this point, I got an idea. I asked Groups 1 and 2 to bring their posters and meet me in the back of the room. I explained that they had approached the problem differently, and I was wondering if they could find a way to relate their two approaches. I thought that Group 1's pairing by two with one left over was similar to the loner number that Group 2 talked about and thought that each group could really help strengthen the argument of the other. I left them to sort things out and moved on to Group 4.

Group 4 had used an approach similar to Group 3's, but instead of using two different odd numbers, they were adding the same odd number to itself (as shown in Figure D.4). I wanted to make sure I challenged them on this before they finished. I asked Charmaine to explain what the group had done so far. She explained, "$2x$ is even, so $2x + 1$ is odd, and if you add it to itself you get three numbers, all of which are divisible by 2." I said, "Suppose that $x = 4$. What are you adding together?" Milan said, "It would be $9 + 9$, which is 18, which is even." I replied, "But in your argument, aren't you always adding the same odd number to itself? How could you show that the conjecture works for two *different* odd numbers?" With that, I left them to continue to ponder the question.

FIGURE D.4 Work produced by students in Group 4.

2x + 1 is odd
2x + 1 + 2x + 1 is adding two odds
2x + 2x + 2 is even because all parts are divisible by 2

Group 4: Jason, Leah, Milan, and Charmaine

At this point, I noticed that less than 10 minutes was left on the timer. I needed to get back to Group 6 and see what they had come up with. I noticed that they had taken my suggestion and had added drawings to the poster (shown in Figure D.5). I asked Mike to explain what they had added to the poster and why. He said, "Well, we decided to take Tamika's idea that odd numbers divided by 2 had one left over and see if we could make a model like you said. We got some square tiles and started with $5 + 3$. We counted out five gray tiles and arranged them into two rows, but the rows didn't have the same number of tiles in them. Then we counted out three white tiles and again made two uneven rows with them. Then we noticed how they would fit together and form a rectangle that was 2 by 4. We did this for $3 + 3$ too, but this is as far as we got." I asked the group to consider whether the models gave them a way to think about what all odd numbers would look like. I decided not to wait for an answer. Time was up and I needed to get ready for the discussion.

   WE REASON & WE PROVE FOR ALL MATHEMATICS

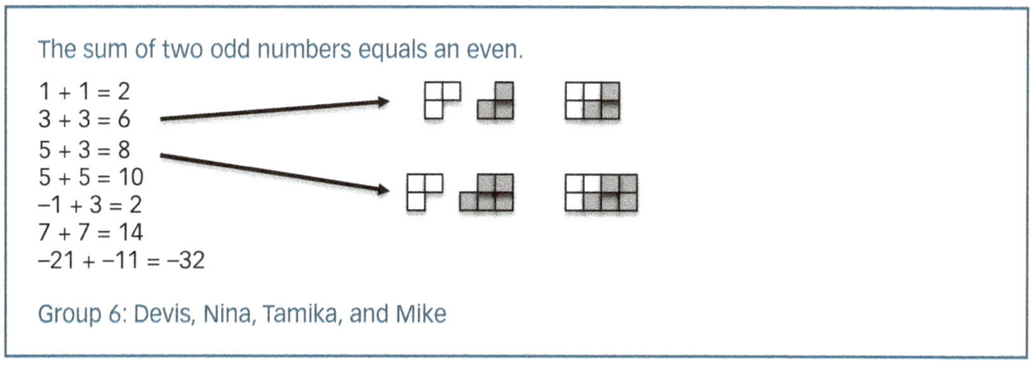

FIGURE D.5 Work produced by students in Group 6.

155 With only a few minutes left before the timer would go off, I quickly reviewed the notes I made as I moved around the room so I could decide the order for discussing the posters. I ultimately wanted everyone to understand the algebraic approach that had been used by Group 3, but I didn't think this was a good place to start since most of the groups did not use algebra. Also, I wanted students to see that the visual approaches that had been used by

160 Groups 1 and 6 could help explain why the conjecture was always true and could be connected to the algebra. I decided to end with the Group 6 poster because by the time we got to it, I figured that students would be able to see why the examples were not sufficient, but also consider the issue of negative numbers that no other group had raised.

165 **FACILITATING WHOLE-GROUP DISCUSSION**

When the timer sounded, I told students to take a minute to finish up and then to bring their posters to the front of the room. I decided to use the front board as the staging area and had written on the top of the board the number of the posters in the order I wanted them placed. The sequence of posters went like this:

170

    Group 1 (dots circled in groups of two)

    Group 2 (general narrative explanation—loner numbers)

    Group 5 (description of the algebra)

    Group 3 (algebra with two different variables)

175   Group 4 (algebra with one variable initially)

    Group 6 (examples plus rectangles)

I started the discussion by asking Group 1 to explain what they had done. Chris, basically repeating what Angela had told me earlier, said, "We decided to show that when you added

180 11 and 13, you really had five sets of 2 for the 11 with one left over and six sets of 2 for the 13 with one left over, and if you add the two leftover ones, you got another group of 2. But Mrs.

E asked if this showed that the sum of any two odd numbers was always even and it really
didn't." At this point, Yolanda jumped in and said, "But we got together with Group 2 and
talked about what they did and then realized that no matter what odd number you have you
185   will also have an extra one left over—the loner number—and that two loner numbers always
make another group of 2." She pointed to the poster produced by Group 2 (shown in Figure
D.6). I asked if anyone had any questions or comments about what Groups 1 and 2 had done
or how the two were connected. Mike (Group 6) asked, "I am not sure what a loner number
is." Jamilla, a member of Group 2, said, "A loner number is really just the 1 that is left over
190   when you divide an odd number by 2. The picture Group 1 drew really shows the loner number 1 and how it gets paired with the other loner number 1 to make an even number."

FIGURE D.6 Work produced by students in Group 2.

> Every odd number has one loner number. Add the two loner numbers and you will get
> an even number. Now add all together the loner numbers and the other two (now even)
> numbers.
>
> Group 2: Jamilla, Amanda, Osmi, and Kalib

I then asked Group 5 how their poster (shown in Figure D.7) related to the two we had
already discussed. Ramon came up to the front of the room to explain, "This statement
(pointing to the Poster 2 statement that *every odd number has one loner number*) is really
195   the same as our even + 1, and this drawing (pointing to the Poster 1 diagram of 11 as five
groups of 2 with one left over) shows even + 1 as a picture."

FIGURE D.7 Work produced by students in Group 5.

> Even # = odd # + odd #
>
> Even # = even # + 1 + even # + 1
>
> Even # = even # + even # + 2
>
> The reason why this equation is true is because all of the values are always evens and
> divisible by 2.
>
> Group 5: Renee, Shakita, Ramon, and Philip

I took a quick look at the time to see that we only had about 10 minutes left in class and
I still had ground to cover. I asked Group 3 to talk about what they did and how it was
connected to the other posters. Ben, referring to the poster his group had produced (shown

    WE REASON & WE PROVE FOR ALL MATHEMATICS

200   in Figure D.8), explained, "Ours really isn't very different, we just used algebra from the beginning. Instead of saying even like Group 5 did, we used $2x$ and $2a$." Osmi (Group 2) was raising his hand at this point. I pointed out to Ben that it looked like someone had a question. Ben called on Osmi, who asked, "How did you know you could write an even number as $2x$?" Ben said, "I remember that we did that last year in algebra when we were

205   working on word problems." Kim, a member of Ben's group, jumped in, "All even numbers are divisible by 2, right? So that means that even numbers must also be multiples of two." I asked Osmi if that made sense and he said, "It does now, but I am not sure I would have thought of it."

FIGURE D.8 Work produced by students in Group 3.

Even number: $2x$

Odd numbers: $2x + 1$

Even number: $2a$

Odd number: $2a + 1$

$2x + 1 + 2a + 1 = 2x + 2a + 2$

Group 3: Louis, Kim, Ben, and Sophia

I had originally thought that we would be able to start the second part of this task (about the

210   product of any two odd numbers) toward the end of class or that I could assign it for homework. I could see now that those plans were not going to work—I felt that the students were not quite ready to leave the sum of two odds part of the task. I had to do some quick thinking, and with only a few minutes left, I told students that for homework, they needed to answer three questions about the two posters we had not yet discussed. I wrote these questions on the board:

215

1. Do you need two *different* variables if you are showing the conjecture is always true using algebra?
2. Does the conjecture that the sum of two odd numbers is even apply to all integers?
3. Does the argument presented on Poster 6 count as a proof? If not, what could you

220   add to it to make it a proof?

I encouraged students to look carefully at the posters from Groups 4 and 6 in order to get a bit more insight on these questions since Group 4 had amended their initial work to include two different variables (as shown in Figure D.9) and Group 6 had used positive and negative integers

225   in their examples, but had only used examples. I told them that I would post pictures of all of the posters on our class website so that they could refer to them when doing their homework.

FIGURE D.9 Work produced by students in Group 4.

FIGURE D.9 Work produced by students in Group 4.

$2x + 1$ is odd

$2x + 1 + 2x + 1$ is adding two odds **(the same odds)**

$2x + 2x + 2$ is even because all parts are divisible by 2

$2y + 1$ is also odd

$2x + 1 + 2y + 1$ is adding two different odd numbers

$2x + 2y + 2$ is even because all parts are divisible by 2

Group 4: Jason, Leah, Milan, and Charmaine

When the bell rang, students quickly packed up their stuff and went off to their next class. I started to think about where I wanted to go with this discussion. I decided to start the class the next day with a discussion of the characteristics of proof and what elements were necessary for an argument to qualify as proof. This would build nicely on the final homework question and would be a supportive precursor to them working on the product of two odd numbers conjecture.

230

Appendix E

Making Sure That All Students Understand: The Case of Calvin Jensen

In his math classes, Calvin Jensen used curriculum materials that provided a lot of opportunities for students to engage in activities related to reasoning-and-proving. The curriculum frequently used contextual problems to engage students in looking for patterns, making conjectures, and creating proof and non-proof arguments. Calvin liked these types of problems
5 *because he felt that the context could really support students' reasoning and sense-making. He selected the Sticky Gum problem early in the year because he thought that students would find it interesting and there were lots of ways to approach the task.*

His goal for this lesson was for students to determine the maximum amount that would be
10 *spent to satisfy the children in each of three situations: twins and two gumball colors, twins and three gumball colors, and triplets and three gumball colors. Once these maximums had been determined, he wanted his students to use what they learned in the three situations to write and justify an algebraic generalization for finding the maximum cost for any number of children and colors.*
15

CALVIN TALKS ABOUT HIS CLASS

Since it's early in the year, I'm excited to see what my students can do with the Sticky Gum problem. They have done some problems that ask them to justify their conclusions, but I have a number of students who still struggle with articulating what they are thinking.
20 Because of this, I didn't want to give them the problem without some scaffolding to get them started. I've used the strategy of "What do we know?" in the past, and I think that it is a good strategy for starting today's lesson.

THE FIRST PART OF CLASS

25 To aid my class in getting started, we spent about 15 minutes reading the problem and discussing and organizing "what we know" about each question. Figure E.1 contains what I wrote on the board based on the whole-class discussion.

What do we know?		
Question 1	**Question 2**	**Question 3**
• Two different colored gumballs in the machine • Two children • Want same color gumball • Gumball costs a penny • Each twin gets one gumball • Answer is 3 cents	• Three different colored gumballs • Two children • Twins still want same color • Gumball costs a penny • Each twin gets one gumball • Probability?	• Triplets • Three different colored gumballs • All kids want same color gumball • Cost is still a penny • One gumball per triplet

I anticipated that students would struggle to fully understand why the answer to the first question is three gumballs rather than two. I wanted to make sure students understood that two gumballs is possible, but having to buy three gumballs is the worst-possible case. I gave each group a cup of colored cubes, which they could use to explain the three scenarios.

MONITORING SMALL GROUP WORK

I checked in on Robert, Tim, Sherrell, and Rochelle's group as they were getting started: "Tell me what you understand about Question 1." Rochelle explained that the twins want gumballs, but the colors have to be the same. I asked, "What is the actual question you are trying to answer? What are you trying to find?" Sherrell softly said, "How much the mother has to spend." Tim quickly said that the gumballs could cost 2 cents or they could cost more. Robert reread the text and added that the cost would be 3 cents. Tim responded to Robert, explaining that it could cost 2 cents if the first two gumballs are the same color, or she could spend more than 3 cents. I interjected, "But all she needs to get is what?" Robert said, "Two of the same color." I then reminded them that there are only two colors and asked, "Could she spend less than 3 cents?" Sherrell nodded her head and turned toward Robert and said, "If the first two are the same color, then it could be 2 cents, but we wrote on the board that the cost is 3 cents." As the group thought about what Sherrell said, I told them that they need to continue their conversation to figure out why three is the answer using a model, the cubes, or some other method.

I moved on to another group (Tia, Devin, and Max) and immediately asked, "What is Question 1 asking?" None of the students had written anything down on paper, but they each had a pile of colored cubes in front of them. Speaking to the group, I said, "Since you have had time to work on the question, what have you concluded? Why is the answer 3 cents?" Devin used the cubes to explain why 3 cents was the worst-case scenario. I pushed him on whether the answer could be 2 cents. He said that 2 cents is possible, but 3 cents is the worst case—when the twins would have the same color gumball. He went on to say that 2 cents would be the least case and 3 cents was the worst case.

The other two students in the group were listening, but when I asked them questions, neither seemed ready to explain the solution. I picked up two red cubes and asked, "What if I put in a penny and I get a red one and put in another penny and get a second red gumball? Are the twins satisfied? They both have the same colored gumball and I only spent 2 cents." Tia wondered aloud, "Is this about probability?" Max explained that it could be since it is the chance of getting 3 cents. I tried to refocus the group to think about what Question 1 was asking, "Does it ask us about percentages or chances or something else?" Devin said, "It's about how much money." So, I took the two red cubes and asked the group again, "I got two of the same colored gumballs and only spent 2 cents, but the answer says 3 cents." Devin looked back at the problem in the book and said that it asks for the most that she would have to pay. I asked him to explain the difference in what I was saying and what he was saying. Devin continued, "Two cents would be the least amount that she could spend and 3 cents is the most that she could spend and they are looking for the most." I replied, "OK, I got it." Before I left the group, I encouraged them to continue to work on the other questions, but to make sure that they worked together so everyone in the group understood.

I then walked over to Tamyra, Merek, Jake, and Jill's group. Tamyra and Merek had both recorded possible combinations of gumballs (Figure E.2), but there were some differences in their combinations.

FIGURE E.2. Tamyra and Merek's work.

Tamyra's Combinations (each column is a different combination)				Merek's Combinations		
1 – W	1 – R	1 – W	1 – R	2 cents	3 cents	3 cents
1 – R	1 – W	1 – W	1 – R	WW (done)	WRW (done)	WRR (done)
1 – W	1 – R	~~1 – R~~	~~1 – W~~	RR (done)	RWR (done)	RWW (done)

I sat down and engaged them in the following conversation:

Calvin: Tell me what you all are doing.

Tamyra: OK, so we understand that you put in a penny and a white one comes out (holding
          her hand about 6 inches above the desk, she drops a white cube). Then you put in
          another penny and a red one comes out (again dropping a red cube to represent a
          red gumball coming out of the machine). Then you put in a third penny, and since
          this one has to be red or white since there are only two colors, it will cost 3 cents.
          But what if you put in a penny and you get a red and then you get a red again?

Merek: You stop. You're done.

Calvin: So if she gets two red or two white gumballs right at the beginning, the answer is
          two, right?

Merek: Yeah.

Calvin: So why are they saying the answer is 3 cents?

Jill: It could cost three or two, but it needs to be three.

Calvin: Well, then, what is the difference between two or three?

Jake: One.

Calvin: There is a difference of one, but what is the difference between the answer being
          two or the answer being three besides the difference of one number?

Tamyra: Like she said (pointing to Jill). You could get two only if the first two are the same
          color. But if they are not the same first two, then it will cost 3 cents.

Calvin: Show me how you get three—the possibility of it costing 3 cents.

Merek: I wrote them out (referring to his work presented in the table).

Calvin: OK, I see. One penny for white and another penny for white is a way it could
          happen. So you did a list of ways it could happen. Right?

Tamyra: I wrote them too, but I got confused. He (Merek) got six and I got four so I don't
          know.

Calvin: What did the rest of you get? OK. I want you guys to work as a group. Share what
          each of you is thinking. Work together. Work as a team. Explain to each other
          what you are doing.

After 40 minutes had passed, the students began to answer Question 2. While I was pleased
to see that they could all explain why 3 cents is the appropriate answer to Question 1, we
only had 20 minutes remaining in the class period. I was left wondering how, at this rate, I
would ever get through the entire problem.

     WE REASON & WE PROVE FOR ALL MATHEMATICS

Appendix F

Helping Students Connect Pictorial and Symbolic Representations: The Case of Natalie Boyer

Natalie Boyer is an experienced teacher who has received many accolades for her teaching. Over the past few years, she has begun to question whether she was providing sufficient opportunities for students who were not in advanced classes to engage with challenging tasks. In an effort to improve this situation, she decided to implement the Sticky Gum prob-
5   *lem, a task she first encountered in a professional development institute that she attended during the previous summer. She thought the problem was a good one for her students since the first three questions would provide scaffolding for students' entry into the problem and would help them in Part 4, which would be more challenging.*

10   *Her goal for the lesson was for students to identify and generalize the relationship between the number of children, the number of colors, and the amount of money that would have to be spent to meet the conditions for each question. She knew that this would not be easy for her students, but felt that with appropriate support, they could meet the challenge the task presented.*
15

NATALIE TALKS ABOUT HER CLASS
My Algebra 2 class has 20 regular education students and 10 learning support students. To make sure *all* students had entry into the Sticky Gum problem, I broke it into three parts: twins with two colors, twins with three colors, and multiple kids with multiple colors. Based
20   on our work on this problem in the professional development institute this summer, I knew that a good strategy was to look for a worst-case scenario for each part. So, I decided I would help the students get started by explaining that scenario.

THE BEGINNING OF CLASS
25   I launched the problem by having the students read through it silently and then asking if there were any questions. Hearing none, I explained the concept of the worst-case scenario. I could see students nodding along, so at the end of my explanation, I put them to work on the problem. As I monitored the students' progress on the problem, I noticed that some groups were having trouble verbalizing the worst-case scenario. I decided to have a few students
30   explain their work on the first two parts of the problem publicly, hoping that this would serve to jump-start the other groups. While different students presented their work to the class, I pressed them to justify their conclusions and also helped them clarify their work with pictures and language choices. As this whole-class presentation progressed, I hoped that

struggling students started to see that their peers who related their justifications to the context of the problem were better able to convince others that their explanations were sound.
Having completed the first two parts of the problem as a whole group, we then moved on to
the third part—multiple kids with multiple colors. This part forced students to form generalizations, and many were confused on how to proceed. To get them started, I suggested to
the whole group that we make a common table of the situations they had already explored
(two kids, two colors and two kids, three colors), then they could generate more data for
the table in their groups until a pattern emerged.

MONITORING SMALL GROUP WORK

As students explored various cases and volunteered ideas about the pattern in their small groups,
I moved from group to group and helped them symbolically represent their patterns and asked
them to record the expressions on the board so that we could discuss them as a group.

While small groups were finishing up and putting their solutions on the board, I made my way
to a quiet group of learning support students in the center of the room. "How is it going?"
I asked. Lee sheepishly pushed his paper to me and remarked that his solution was not any
good. However, I quickly saw that Lee had generated a good amount of data, organized it in a
very useful way, and clearly showed the worst-case scenario (Figure F.1). I asked Lee to explain
what he had done so I could fully understand his thinking. Using a diagram that he created for
10 colors and four kids, Lee shared his thinking: "The letters represent the different gumball
colors that could come out of the machine on each draw except for the last column. Those
gumballs haven't been drawn yet, but any draw from the last column (any color) would satisfy
the four kids because three of the same colors had already been drawn."

FIGURE F.1 Lee's work.

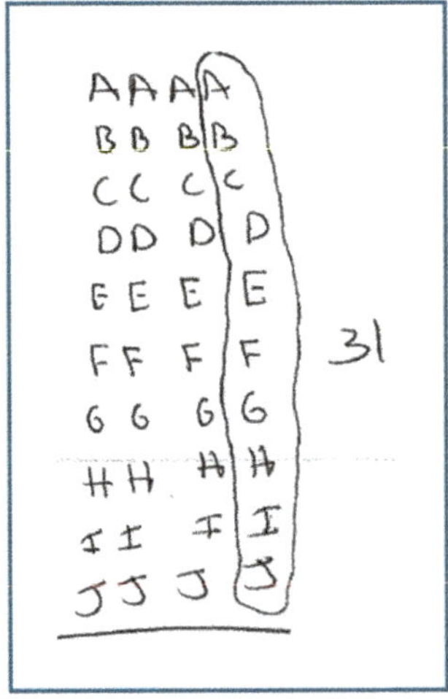

WE REASON & WE PROVE FOR ALL MATHEMATICS

Since Lee had shown 30 gumballs being drawn (A–J in the first three columns) and he knew only one more gumball of any color was required, he calculated an answer of 31. I was excited

60 and said I would help Lee and his group convert their pattern into a formula. I asked Lee what the letters in each column represented and he indicated that each letter stood for one of the ten colors of gumballs. The group agreed to use the notation on which the class had decided (c for number of colors and k for number of kids), and I wrote c as the number of different colors in the first column (see Figure F.2). With some prompting, Lee wrote c at the bottom of the next

65 two columns, and a "1" at the bottom of the fourth column. I placed addition signs between the terms ($c + c + c + 1$), then turned to one of Lee's partners. "Andy, what do you think? Will there always be a 'plus one' on the outside?" Andy indicated that there would be, but when I asked him for a justification, he could not say why he agreed. "Why not plus 2? Or minus 1? What is it about this *specific* problem that makes it 'plus 1'?" I asked. The third partner, Lynn,

70 volunteered that no matter what color you got on the fourth draw it would be a match.

FIGURE F.2 Lee's modified work.

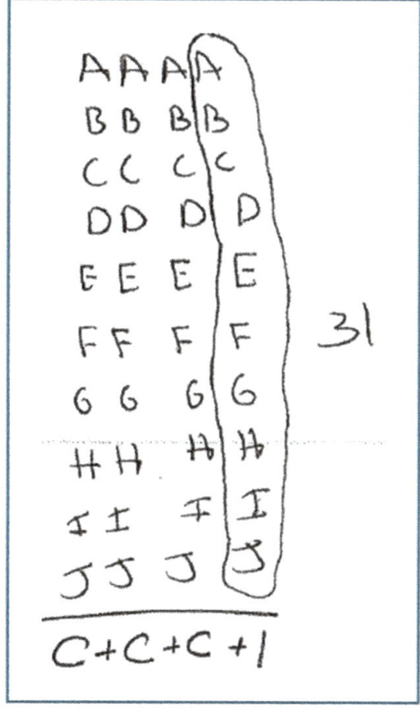

I then pointed out to Lee that his formula did not contain anything about the number of
kids. Lee suggested dividing something by 4 or adding 4, but he could not make the formula
work for his data. I decided to approach this from a different direction and asked Lee and
his partners how they could write "$c + c + c$." Lynn said that they could write it as "$3c$"
and noted that this was one less than the number of kids times the number of colors. She
suggested $(k - 1)c + 1$ as the expression. I asked Lynn to put this expression on the board
and left Lee, Andy, and Lynn to test their expression and explain why it worked in words. I
glanced at the clock and realized that I had been talking to this group for about 5 minutes
and that the bell would ring in about 10 minutes.

FACILITATING WHOLE-CLASS DISCUSSION

I gathered the class together to discuss the three proposed formulas on the board:

- $k + c - 1$
- $kc - 1$
- $(k - 1)c + 1$

I asked Marco to pick a number of colors and a number of kids to work with, and he picked
four colors and six kids. I then asked the students to guide me in writing up a systematic list
for finding the worst-case scenario possible for four colors and six kids. Sandi suggested that
I list one gumball of each color, and I began using an organizational scheme similar to Lee's
column scheme but with two important changes: the letters corresponded to the first letter of
the color instead of the consecutive letters of the alphabet, and the colors were listed in rows
not columns (to reinforce the idea of randomness). I wrote row after row of colors (R, W, G,
and B), asking the students to tell me when to stop (Figure F.3).

FIGURE F.3 Recording of
colors on the board.

R	W	G	B
R	W	G	B
R	W	G	B
R	W	G	B
R	W	G	B
R			

WE REASON & WE PROVE FOR ALL MATHEMATICS

As I started the sixth row, I slowed down and students started yelling "stop" and "keep going." The students saying "stop" convinced the others that only one color was needed in the final row. At this point, Christopher volunteered that he also had a formula: $ck - c + 1$, so I added that to the board as well. I noticed that Christopher's formula was the same as Lynn's formula with the number of colors distributed across the $(k - 1)$ part. I pointed this out, "So it looks like we have two formulas that say the same thing in different ways. Lynn has a formula based on Lee's work $(k - 1)c + 1$, and Christian has a formula based on Erica's work $ck - c + 1$. How do these formulas work, based on this (I pointed to the table of organized colors on the board) picture?"

My question was met with silence. "OK, let's pick Lee's formula apart. What's at the end of your formula?" In relating specific parts of the formula to specific aspects of the picture (which related to the context of the problem), I helped the students see that the "plus 1" in the formula matched the one gumball in the final row and also why the number of colors was multiplied by one less than the number of kids.

Just as we were finishing up this explanation, the bell rang and students hurried from the room. I felt that we had made good progress. I hoped that I hadn't provided too much support to my students. If you provide too little help, students get frustrated, and if you provide too much help, students end up doing too little of the hard thinking. I went to lunch wondering whether I struck the right balance.

References

American Diploma Project. (2004). *Ready or not: Creating a high school diploma that counts.* Available from http://www.achieve.org

Boston, M. D., Dillon, F., Smith, M. S., & Miller, S. (2017). *Taking action: Implementing effective mathematics teaching practices in grades 9–12.* Reston, VA: National Council of Teachers of Mathematics.

Boyle, J. D. (2012). *A study of prospective secondary mathematics teachers' evolving understanding of reasoning-and-proving* (Doctoral dissertation). Available from ProQuest Dissertations and Theses database. (UMI No. 3532776)

Coe, R., & Ruthven, K. (1994). Proof practices and constructs of advanced mathematics students. *British Educational Research Journal, 20,* 41–53.

Coxford, A. F., Fey, J. T., Hirsch, C. R., Schoen, H. L., Burrill, G., Hart, E. W., et al. (2003). *Contemporary mathematics in context: A unified approach: Course 3.* New York: Glencoe McGraw-Hill.

Ellis, A. (2007). Connections between generalizing and justifying: Students' reasoning with linear relationships. *Journal for Research in Mathematics Education, 38,* 194–229.

Ersen, Z. B. (2016). Preservice mathematics teachers' metaphorical perceptions towards proof and proving. *International Education Studies, 9*(7), 88–97. doi:10.5539/ies.v9n7p88

Fendel, D., Resek, D., Alper, L., & Frazer, S. (1996). *Interactive mathematics program year 1 – Unit 2: The book of pig.* Emeryville, CA: Key Curriculum Press.

Fey, J. T., Hirsch, C. R., Hart, E. W., Schoen, H. L., & Watkins, A. E. (2009). *Core-plus mathematics: Contemporary mathematics in context, course 3* (2nd ed.). Columbus, OH: Glencoe/McGraw-Hill.

Freeburn, B. (2015). *Preservice secondary mathematics teachers learning of purposeful questioning and judicious telling that supports students' mathematical thinking* (Unpublished doctoral dissertation). The Pennsylvania State University, University Park, PA.

Freeburn, B., & Arbaugh, F. (2017). Supporting productive struggle with communication moves. *Mathematics Teacher 111,* 176–181.

Goetting, M. (1995). *The college students' understanding of mathematical proof.* Unpublished doctoral dissertation, University of Maryland, College Park.

Gravemeijer, K., & Doorman, M. (1999). Context problems in realistic mathematics education: A calculus course as an example. *Educational Studies in Mathematics, 39,* 111–129.

Harel, G. (2013). Intellectual need. In K. Leatham (Ed.), *Vital directions for mathematics education research* (pp. 119–151). New York: Springer.

Harel, G., & Sowder, L. (1998). Students' proof schemes: Results from exploratory studies. In A. H. Schoenfeld, J. Kaput, & E. Dubinsky (Eds.), *Research in collegiate mathematics education III* (pp. 234–283). Providence, RI: American Mathematical Society.

Harel, G., & Sowder, L. (2007). Toward comprehensive perspectives on the learning and teaching of proof. In F. K. Lester (Ed.), *Second handbook of research on mathematics teaching and learning* (pp. 805–842). Greenwich, CT: Information Age Publishing.

Healy, L., & Hoyles, C. (2000). A study of proof conceptions in algebra. *Journal for Research in Mathematics Education, 31*, 396–428.

Horizon Research, Inc. (2013). *2012 National survey of science and mathematics education: Highlights Report.* Chapel Hill, NC: Author.

Huinker, D., & Bill, V. (2017). *Taking action: Implementing effective mathematics teaching practices in grades K–5.* Reston, VA: National Council of Teachers of Mathematics.

Johnson, G. J., Thompson, D. R., & Senk, S. L. (2010). Proof-related reasoning in high school textbooks. *Mathematics Teacher, 103*, 410–418.

Karunakaran, S., Freeburn, B., Konuk, N., & Arbaugh, F. (2014). Improving preservice secondary mathematics teachers' capacity with generic example proofs. *Mathematics Teacher Educator, 2*, 158–170.

Knuth, E. J. (2002a). Proof as a tool for learning mathematics. *Mathematics Teacher, 95*, 486–490.

Knuth, E. J. (2002b). Secondary school mathematics teachers' conceptions of proof. *Journal for Research in Mathematics Education, 33*, 379–405.

Knuth, E. J., Choppin, J., Slaughter, M., & Sutherland, J. (2002). Mapping the conceptual terrain of middle school students' competencies in justifying and proving. In D. S. Mewborn, P. Sztajn, D. Y. White, H. G. Weigel, R. L. Bryant, & K. Nooney (Eds.), *Proceedings of the 24th annual meeting of the North American chapter of the International Group for the Psychology of Mathematics Education* (Vol. 4, pp. 1693–1700). Athens, GA: Clearinghouse for Science, Mathematics, and Environmental Education.

Lakatos, I. (1976). *Proofs and refutations: The logic of mathematical discovery.* Cambridge, UK: Cambridge University Press.

Lappan, G., Fey, J. T., Fitzgerald, W. M., Friel, S. N., & Philips, E. D. (1998/2004). *Connected mathematics project.* Menlo Park, CA: Dale Seymour Publications.

Lesh, R., Post, T., & Behr, M. (1987). Representations and translations among representations in mathematics learning and problem solving. In C. Janvier (Ed.), *Problems of representation in the teaching and learning of mathematics* (pp. 33–40). Hillsdale, NJ: Erlbaum.

Lloyd, G., Herbel-Eisenmann, B., & Star, J. R. (2011). *Developing essential understanding of expressions, equations, and function for teaching mathematics in grades 6–8.* Reston, VA: National Council of Teachers of Mathematics.

McClure, L., Woodham, L., & Borthwick, A. (2011). Using low threshold high ceiling tasks. Retrieved from http://nrich.maths.org/7701

National Center for Education Statistics. (2003). *Teaching mathematics in seven countries: Results from the TIMSS video study.* Washington, DC: U.S. Department of Education.

National Council of Teachers of Mathematics (NCTM). (1989). *Curriculum & evaluation standards for school mathematics.* Reston, VA: Author.

National Council of Teachers of Mathematics (NCTM). (1991). *Professional standards for teaching mathematics.* Reston, VA: Author.

National Council of Teachers of Mathematics (NCTM). (1995). *Assessment Standards for School Mathematics.* Reston, VA: Author.

National Council of Teachers of Mathematics (NCTM). (2000). *Principles and standards for school mathematics.* Reston, VA: Author.

National Council of Teachers of Mathematics (NCTM). (2006). *Curriculum focal points for prekindergarten through grade 8 mathematics: A quest for coherence.* Reston, VA: Author.

National Council of Teachers of Mathematics (NCTM). (2009). *Focus in high school mathematics: Reasoning and sense making.* Reston, VA: Author.

National Council of Teachers of Mathematics (NCTM). (2014). *Principles to actions: Ensuring mathematical success for all.* Reston, VA: Author.

National Governors Association Center for Best Practices, Council of Chief State School Officers (NGA, CCSSO). (2010). *Common core state standards – mathematics.* Washington D.C.: Author.

Nelson, R. B. (1997). *Proof without words: Exercises in visual thinking.* Washington, DC: Mathematical Association of America.

Reid, D. A. (2002). Conjectures and refutations in grade 5 mathematics. *Journal for Research in Mathematics Education, 33,* 5–29.

Schoenfeld, A. (1994). What do we know about Mathematics Curricula? *Journal of Mathematical Behavior, 13*(1), 55–80.

Smith, M. S. (2001). *Practice-based professional development for teachers of mathematics.* Reston, VA: National Council of Teachers of Mathematics.

Smith, M. S., Bill, V., & Hughes, E. K. (2008). Thinking through a lesson protocol: A key for successfully implementing high-level tasks. *Mathematics Teaching in the Middle School, 14,* 132–138.

Smith, M. S., Steele, M. D., & Raith, M. L. (2017). *Taking action: Implementing effective mathematics teaching practices in grades 6–8.* Reston, VA: National Council of Teachers of Mathematics.

Smith, M. S., & Stein, M. K. (1998). Selecting and creating mathematical tasks: From research to practice. *Mathematics Teaching in the Middle School, 3,* 344–350.

Smith, M. S., & Stein, M. K. (2018). *5 practices for orchestrating productive mathematics discussions, Second Edition.* Reston, VA: National Council of Teachers of Mathematics.

Sowder, L., & Harel, G. (2003). Case studies of mathematics majors' proof understanding, production, and appreciation. *Canadian Journal of Science, Mathematics and Technology Education, 3,* 251–267.

Stein, M. K., Grover, B. W., & Henningsen, M. (1996). Building student capacity for mathematical thinking and reasoning: An analysis of mathematical tasks used in reform classrooms. *American Educational Research Journal, 33,* 455–488.

Stigler, J. W., & Hiebert, J. (2004). Improving mathematics teaching. *Educational Leadership, 61*(5), 12–16.

Stylianides, A. J. (2007). Proof and proving in school mathematics. *Journal for Research in Mathematics Education, 38,* 289–321.

Stylianides, A. J. (2009). Breaking the equation "empirical argument = proof." *Mathematics Teaching, 213,* 9–14.

Stylianides, A. J., & Stylianides, G. J. (2009). Proof constructions and evaluations. *Educational Studies in Mathematics, 72*(2), 237–253.

Stylianides, G. J. (2008a). An analytic framework of reasoning-and-proving. *For the Learning of Mathematics, 28*(1), 9–16.

Stylianides, G. J. (2008b). Proof in school mathematics curriculum: A historical perspective. *Mediterranean Journal for Research in Mathematics Education, 7*(1), 23–50.

Stylianides, G. J. (2009). Reasoning-and-proving in school mathematics textbooks. *Mathematical Thinking and Learning, 11*(4), 258–288.

Stylianides, G. J. (2010). Engaging secondary students in reasoning and proving. *Mathematics Teaching, 219,* 39–44.

Stylianides, G. J., & Stylianides, A. J. (2009). Facilitating the transition from empirical arguments to proof. *Journal for Research in Mathematics Education, 40,* 314–352.

Weiss, I. R., Pasley, J. D., Smith, P. S., Banilower, E. R., & Heck, D. J. (2003). Looking inside the classroom: A study of K–12 mathematics and science education in the United States. Available online at http://www.horizon-research.com/insidetheclassroom/reports/looking/

Index

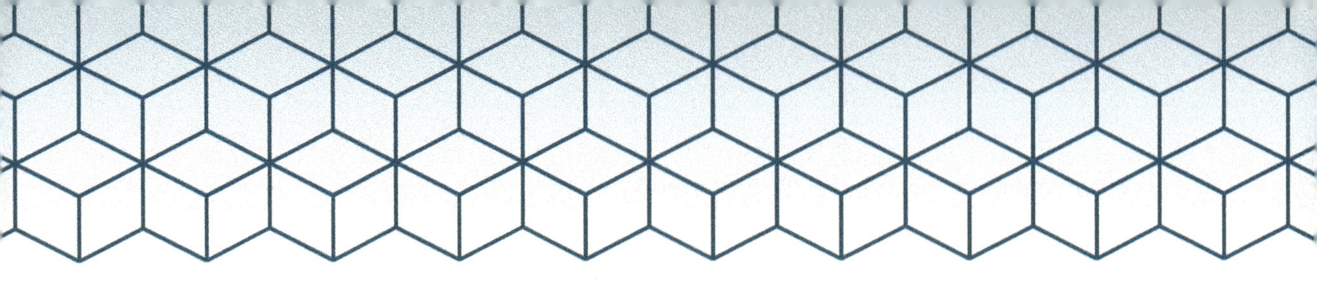

Notes

Notes

Notes

Notes

Notes

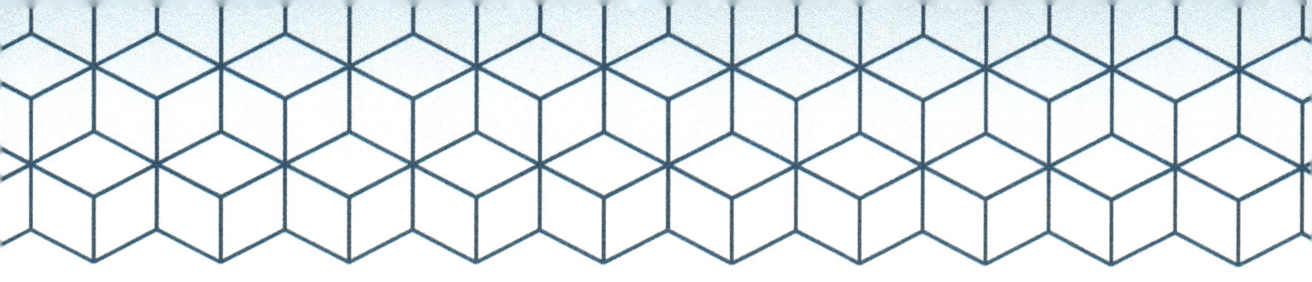

Notes

Notes

Notes